Robust Situation Awareness in Tactical Mobile Ad Hoc Networks

Vom Fachbereich Informatik
der Technischen Universität Darmstadt
genehmigte

DISSERTATION

zur Erlangung des akademischen Grades eines
Doktor-Ingenieurs (Dr.-Ing.)
von

Dipl.-Inform. Peter Ebinger

geboren in Backnang

Referenten der Arbeit:	Prof. Dr. techn. Dieter W. Fellner Technische Universität Darmstadt
	Prof. Dr.-Ing. Stephen D. Wolthusen Royal Holloway, University of London, UK
Tag der Einreichung:	29. Juli 2013
Tag der mündlichen Prüfung:	9. September 2013

Darmstädter Dissertationen 2013
D 17

Bibliografische Information der Deutschen Nationalbibliothek

Die Deutsche Nationalbibliothek verzeichnet diese Publikation in der Deutschen Nationalbibliografie; detaillierte bibliografische Daten sind im Internet über http://dnb.d-nb.de abrufbar.

ISBN 978-3-8325-3560-5

Logos Verlag Berlin GmbH
Comeniushof, Gubener Str. 47,
10243 Berlin
Tel.: +49 (0)30 42 85 10 90
Fax: +49 (0)30 42 85 10 92
INTERNET: http://www.logos-verlag.de

Acknowledgements

I would like to express my greatest gratitude to the people who have helped and supported me throughout my dissertation project. First of all, I wish to express my thanks to my principal supervisor Prof. Dieter W. Fellner for providing me the opportunity to implement my dissertation at Fraunhofer IGD. His help and support made this thesis possible.

I am highly indebted to my former colleagues at the Security Technology (now Identification and Biometrics) department at Fraunhofer IGD for the inspiring, positive and productive working atmosphere and for providing the infrastructure to finish my dissertation project even after leaving Fraunhofer IGD. I would like to express my great appreciation of the personal support, the discussions and constructive feedback to Prof. Christoph Busch, Dr. Arjan Kuijper, Alexander Nouak, Alexander Opel, Jan Peters, Dr. Ulrich Pinsdorf, Dr. Martin Schmucker, Helmut Seibert and Dr. Xuebing Zhou.

I am grateful to my students for the numerous discussions and their help to implement and validate important buildings block of my dissertation project. I would like to particularly thank Stefan Appel, Norbert Bißmeyer, Fatih Gey, Malcolm Parsons, Dr. Steffen Reidt, and Martin Sommer.

My very special thanks are dedicated to Prof. Stephen D. Wolthusen, my second supervisor. Without his continuous support and his valuable and constructive suggestions this research work would not have been possible. He accompanied me through the whole dissertation project, from initial advice and project acquisition in the early stages through ongoing advice and encouragement to this day.

Finally, I wish to thank all of my family for their support and encouragement throughout my dissertation project. Above all, I would like to thank my wife Julia Ebinger for her personal support and great patience during these times and our children Klara and Lukas Ebinger for being so loving and patient.

Abstract

The objective of the research presented in the dissertation at hand is to improve situation awareness in tactical mobile ad hoc networks (MANETs). Tactical teams are supported to successfully accomplish their missions. This could be a team of first responders after a natural catastrophe or a terrorist attack. For example, after the Haiti earthquake in 2010, rescue forces provided technical and humanitarian aid for the affected population. Small teams tried to find victims and provide food and water supplies in difficult to access areas on the island. In this kind of scenarios, the situation is dynamically evolving and robust and effective mechanisms are required to enable task forces to quickly respond to recent developments. It is crucial that they are aware of the current situation, e. g., location and status of task forces and victims, resources and other entities such as supplies and medication. This information is crucial for mission success as it enables informed decisions and can save lives. The existing infrastructure may be destroyed (or should not be used for other reasons) and a backup communication infrastructure needs to be set up. Mobile devices connected by a MANET are a quick, flexible and efficient way to provide such an infrastructure.

Task forces need to be provided with adequate mobile applications for team and resource management, victim support and situation assessment, e. g., location of victims, team situation, presence of harmful substances or buildings in danger of collapse. Reliable and accurate state estimation is the basis technology for effective situation awareness applications. The primary objective explored in the dissertation at hand is increasing the robustness of state estimation. However, new threats and challenges to IT systems in tactical environments arise due to technological advances and ongoing developments. Cyber attacks are easier to implement as attack toolkits are freely available online to the general public or provided as a commodity to organized cybercrime. After a successful attack, terrorists may attack again, targeting rescue workers and IT systems in order to increase the extent of the damage and the number of casualties.

Existing situation awareness solutions for tactical MANETs are not sufficiently protected against faults and security threats. The proposed approaches for data processing and situation assessment are limited in dealing with wrong (malicious or faulty) information. In distributed environments, such as tactical MANETs, collaborative aspects are important for the cooperative information assessment and trustworthiness of remote information sources.

We provide *three main contributions* within this dissertation in order to increase the robustness of situation awareness in tactical MANETs: *cross data analysis*, *cooperative trust assessment* and *probabilistic state modeling*.

The *cross data analysis* concept provides a framework for exploiting all available data sources. The proposed concept incorporates *direct and indirect sensor data and additional knowledge sources* (such as mission-specific knowledge and general information sources) in order to detect inconsistencies. For this purpose, we analyze overlaps and cross-relationships between data sources. *Multiple data sources are analyzed collectively* and cross-relationships are exploited to detect malicious data. We discuss the terms consistency, inconsistency and contradictions within this context. The explicit *modeling of cross-relationships* within our approach enables the derivation of consistency checks in order to detect false or malicious data. The developed concept for *cross data analysis is applied and implemented as a proof of concept for location related data sources.* Cross data analysis can be utilized to increase robustness of situation awareness in tactical MANETs. Contradictions in data sets can be more effectively detected exploiting cross-relationships between data sources. Cross data analysis helps to cope with wrong data due to sensor faults as well as with malicious behavior of attacking nodes. Detection capabilities for malicious behavior are improved and detection rates are increased. Tactical teams get a clearer picture of the situation and are able to successfully complete their missions.

Provisioning of efficient and robust mechanisms for *cooperative trust assessment* in tactical MANETs is the second main contribution. The focus of the proposed concepts is on the assessment of the trustworthiness of other network nodes which provide information for situation awareness. For this purpose, we *combine efficient broadcast gossiping trust aggregation* and *trust information modeling as (trust, confidence) value pairs*. The presented trust assessment architecture significantly reduces communication and processing overhead exploiting characteristics of the wireless medium. Trust information is distributed to all direct neighbor nodes at the same time and efficient local trust aggregation mechanisms limit the amount of data that is flooded into the network. The proposed concept is implemented including trust calculation for indirect assessments and combination of multiple opinions. The proposed cooperative trust assessment scheme allows a benign majority of nodes to prevail and accurately classify network nodes based on observations and trust estimations. The resistance against attackers is increased as nodes responsible for wrong or malicious behavior can be identified and excluded from the situation awareness process.

The third contribution is the *probabilistic state modeling and estimation* concept for situation awareness in tactical MANET. It increases the overall robustness regarding error prone or malicious input data. The proposed *concept based on particle filters* is open and *applicable to a wide range of application scenarios.* This includes nonlinear state propagation, Non-Gaussian process noise as well as hybrid state estimation comprising discrete and continuous state variables. Within the proposed state estimation concept we *incorporate multiple observations and additional information sources*, e.ġ., mission information, domain-specific background knowledge and general information sources. These information sources may pose some constraints on the state estimation process and are exploited to adjust the likelihood of specific system states or exclude impossible states. We present a *distributed state estimation architecture* addressing the particular challenges of tactical MANETs. Suitable observations and state estimates are selected and exchanged within tactical MANETs for improved cooperative state estimation. We propose a re-simulation mechanism for resolving temporal and causal correlation of observations and estimations performed by other network nodes. The concepts are applied to *Task Force Tracking*, a typical *example application for tactical situation awareness.* Probabilistic state modeling increases the robustness of the state estimation process to measurement noise. Incorporating additional information sources improves the preciseness of state estimation based on error-prone data sources. State estimation results are more accurate and the situation awareness picture is more realistic. This way, tactical task forces can be supported as good as possible in order to successfully complete their missions.

Zusammenfassung

Ziel der Forschung dieser Dissertation ist die Verbesserung der Situation Awareness in taktischen mobilen Ad-hoc-Netzen (MANETs). Taktische Einsatzkräfte sollen unterstützt werden, um ihre Mission erfolgreich zu erfüllen. Dies kann z. B. ein Team von Ersthelfern nach einer Naturkatastrophe oder einem Terroranschlag sein. In solchen Szenarios entwickelt sich die Situation dynamisch und es werden robuste und effektive Mechanismen benötigt, damit Einsatzkräfte schnell auf aktuelle Entwicklungen reagieren können.

Naturkatastrophen (z. B. ein Tsunami oder ein Wirbelsturm) oder von Menschen verursachte Katastrophen (z. B. eine Explosion in einer Chemiefabrik oder ein terroristischer Anschlag) haben oft erhebliche Auswirkungen auf die Bevölkerung sowie die Infrastruktur in den betroffenen Gebieten. Das Ziel der Rettungs- und Notfallkräfte (wie z. B. Technisches Hilfswerk, THW) ist es, in solchen Situationen technische und humanitäre Hilfe bereitzustellen. Rettungs- und Notfallkräfte müssen in der Regel auf eine temporäre Infrastruktur zurückgreifen bis die bestehende Infrastruktur wieder aufgebaut ist und in Betrieb genommen werden kann. Zum Beispiel leisteten nach dem Erdbeben auf Haiti im Januar 2010 Rettungskräfte technische und humanitäre Hilfe in den betroffenen Gebieten. Kleine taktische Teams suchten auf der Insel in schwer zugänglichen Gebieten nach Opfern. Ihre Mission war besonders schwierig, da die vorhandene Infrastruktur zerstört war.

Die geeignete, effiziente und effektive Unterstützung durch Informationstechnologie (IT) ist entscheidend für den Erfolg solcher Missionen. Den Einsatzkräften müssen geeignete mobile Anwendungen für das Team- und Ressourcen-Management, zur Betreuung und Versorgung der Opfer und zur Lagebeurteilung zur Verfügung gestellt werden. Diese Werkzeuge sind die Basis für umfassende Situation Awareness und eine zielorientierte Command&Control-Infrastruktur. Entscheidend ist, dass sie sich der aktuellen Situation bewusst sind, d. h. dass sie z. B. Position und Status der Einsatzkräfte, der Opfer, der Ausrüstung und anderer Ressourcen kennen und ebenfalls über mögliche Gefahren informiert sind, wie z. B. vorhandene Gefahrenstoffe und einsturzgefährdete Gebäude. Diese Informationen sind von entscheidender Bedeutung für den Erfolg der Mission, da sie gut fundierte Entscheidungen ermöglichen und so Leben retten können.

Zuverlässige und präzise Zustandsschätzung (engl. situation assessment) ist die Basis für effektive Situation-Awareness-Anwendungen. Primäres Ziel dieser Dissertation ist es die Robustheit der Zustandsschätzung zu erhöhen. Bei den beschriebenen Szenarios ist typischerweise die Kommunikationsinfrastruktur zerstört (oder soll aus anderen Gründen nicht verwendet werden) und es muss eine Backup-Kommunikationsinfrastruktur eingerichtet werden. Mobile Geräte, die zu einem MANET verbunden werden, bieten eine schnelle, flexible und effiziente Möglichkeit, eine solche Infrastruktur ad hoc aufzubauen.

Durch den technologische Fortschritt und andere Entwicklungen und Veränderungen entstehen neue Bedrohungen und Herausforderungen für IT-Systeme. Cyber-Angriffe sind leichter zu implementieren, da die benötigten Werkzeuge einer breiten Öffentlichkeit online frei zur Verfügung stehen oder als Ware für die organisierte Cyberkriminalität bereitgestellt werden. Nach einem Anschlag werden möglicherweise auch Rettungskräfte und IT-Systeme von Terroristen angegriffen, um das Ausmaß der Schäden und die Zahl der Verletzten zu erhöhen. Ihr Ziel ist es dann, den regulären Betrieb auf alle denkbaren Weisen stören. Daher ist auch bei Back-up-Infrastrukturen die Robustheit und Ausfallsicherheit von zunehmender Bedeutung.

Situation-Awareness-Lösungen für taktische MANETs sind bisher nicht ausreichend gegen Störungen und Sicherheitsrisiken geschützt. Bestehende Ansätze für die Informationsaufbereitung und Lagebeurteilung können nur sehr eingeschränkt mit falschen (bösartigen oder fehlerhaften) Informationen umgehen. Sie verarbeiten Daten aus verschiedenen Datenquellen separat und schöpfen das Potential von korrelierten oder redundanten Datenquellen nicht aus. Widersprüche innerhalb von Datensätzen aufgrund fehlender oder fehlerhafter Information fallen u. U. auf, es wird aber nicht zusammen mit anderen verfügbaren Datenquellen versucht ein möglicherweise bösartiges Verhalten zu erkennen und diesem entgegenzuwirken.

In verteilten Umgebungen wie taktischen MANETs spielen kooperative Aspekte bei der gemeinsamen Auswertung von Informationen und der Bewertung der Vertrauenswürdigkeit von Informationsquellen eine wichtige Rolle. Das betrifft insbesondere die Möglichkeiten, die sich aus den speziellen Eigenschaften der drahtlosen Kommunikation ergeben, z. B. die gleichzeitige Verteilung von Informationen an alle direkten Nachbarknoten. Weiterhin werden die zeitlichen und kausalen Zusammenhänge bei der verteilten Zustandsschätzung und die damit verbundenen Problemstellungen bisher nicht ausreichend berücksichtigt und gelöst.

Das Ziel dieser Arbeit ist es, diese Defizite bestehender Lösungen zu adressieren. Die Mobilität der Netzwerkknoten und damit verbundene Aspekte stellen dabei eine zentrale Herausforderung in MANET-Umgebungen. Daher sind mobilitäts- und ortsbezogene Aspekte ein Schwerpunkt dieser Dissertation. Das übergeordnete Ziel ist es, die Robustheit der Zustandsschätzung für Situation Awareness in taktischen MANETs zu erhöhen.

Wir adressieren dieses Ziel anhand der folgenden drei Forschungsfragen:

1. Wie können ortsbezogene *Datenquellen* in einer *umfassenden und effektiven Weise* genutzt werden, um bösartiges Verhalten zu erkennen?
2. Wie kann die *Vertrauenswürdigkeit* anderer Netzwerkknoten *effizient und kooperativ* beurteilt werden?
3. Wie können *Systemzustände* auf eine *allgemein anwendbare Weise modelliert* und *umfassend ausgewertet* werden, um gegen falsche und fehlerhafte Daten resistent zu sein?

Im folgenden Abschnitt fassen wir unsere Forschungsbeiträge in Bezug auf die drei identifizierten Forschungsfragen zusammen.

(1) Wie können ortsbezogene *Datenquellen* in einer *umfassenden und effektiven Weise* genutzt werden, um bösartiges Verhalten zu erkennen? Um die Robustheit von Situation Awareness in taktischen MANETs zu erhöhen, entwickeln wir zunächst ein generelles Konzept zur Cross-Data-Analyse. Das vorgestellte Konzept ermöglicht es die verfügbaren Datenquellen umfassend zu nutzen, um bösartiges Verhalten zu erkennen. Dies beinhaltet sowohl direkte und indirekte Sensordaten als auch zusätzliche Informationsquellen (z. B. missionsspezifische Daten und allgemeine Datenquellen), um Unstimmigkeiten zu erkennen.

Zu diesem Zweck analysieren und modellieren wir Überschneidungen und Querbeziehungen zwischen Datenquellen. Zur Erfassung und Analyse sich überschneidendender Datensätze ist neben der räumlichen und zeitlichen Korrelation der Datenelemente auch der Gegenstand (Entität) zu berücksichtigen, auf das sie sich beziehen. Dabei werden mehrere Datenquellen zusammen analysiert und bestehende Querbeziehungen genutzt, um falsche und/oder fehlerhafte Daten zu erkennen.

Zunächst wird eine Analyse aller Datenquellen durchgeführt, die für Situation Awareness in taktischen MANETs zur Verfügung stehen. Das beinhaltet direkte und indirekte Sensordaten, missionsspezifische und allgemeinen Datenquellen. Mögliche Überschneidungen dieser Datenquellen und deren Querbeziehungen werden untersucht und kategorisiert. Mögliche Definitionen der Begriffe Inkonsistenz und Konsistenz von Datenquellen werden analysiert und ein Konzept für die Modellierung von Inkonsistenzen erstellt.

Wir untersuchen, welche Art von Querbeziehungen für die Cross-Data-Analyse genutzt werden können und kategorisieren diese gemäß ihrer Eigenschaften. Wir leiten Definitionen für die Begriffe Konsistenz, Inkonsistenz und Widerspruch in diesem Kontext ab. Die explizite Modellierung von Querbeziehungen in unserem Ansatz ermöglicht die Ableitung von Konsistenzprüfungen. Diese Konsistenzprüfungen können genutzt werden um Widersprüche zwischen mehreren Datensätzen zu entdecken, und so falsche oder schädliche Daten zu erkennen. Das entwickelte Konzept erhöht die Robustheit von Situation Awareness in taktischen MANETs, da nicht nur Inkonsistenzen innerhalb einer Datenquelle erkannt werden, sondern auch Querbeziehungen zwischen mehreren Datenquellen genutzt werden, um Inkonsistenzen zu entdecken.

Das entwickelte Konzept der Cross-Data-Analyse wird als Proof of Concept auf ortsbezogene Datenquellen angewendet. Dazu werden die in taktischen MANETs verfügbaren ortsbezogenen Datenquellen und ihre Querbeziehungen in konsistenter Weise modelliert und analysiert. Insbesondere bestehen Querbeziehungen zwischen der Netzwerktopologie und den Knotenpositionen. Es wird ein effizientes topographiebasiertes Funkausbreitungsmodell entwickelt und implementiert, um die Topographie der Umgebung als zusätzliche Informationsquelle zu nutzen. So können Funksignalstärkemessungen als indirekte Sensordaten für die Abstandsschätzung ausgewertet werden. So lassen sich Knotenpositionen, Routing-Informationen, Funksignalmessungen und ein Topographie-Modell der Umgebung zusammen für die Cross-Data-Analyse nutzen. Wir entwickeln Algorithmen für die Erkennung von Inkonsistenzen und die Detektion von aktiven Angriffen in taktischen MANETs basierend auf Konsistenzprüfungen zwischen den verfügbaren Datensätzen.

Das vorgestellte Cross-Data-Analyse-Konzept wird in Feldversuchen und mit Hilfe von Simulationsumgebungen evaluiert. Feldversuche in verschiedenen Evaluierungsszenarios zeigen, dass Funksignalstärke generell für die Schätzung von Knotenentfernungen genutzt werden können. Die Präzision und Anwendbarkeit hängt allerdings stark vom spezifischen Einsatzszenario ab. Das topographiebasierte Funkausbreitungsmodell wird als Erweiterung des Netzwerk-Simulators ns-2 und des Visualisierungstools iNSpect implementiert. Das Modell zur Berechnung der Reflexions- und Beugungsfaktoren des topographiebasierten Funkausbreitungsmodells wird durch Simulation analysiert und verifiziert. Die simulative Auswertung zeigt, dass die Eigenschaften des topographiebasierten Funkausbreitungsmodells für alle Evaluierungsszenarios konsistent zu bestehenden Konzepten aus der Literatur sind. Die Evaluierung zeigt, dass das implementierte Modell für den Echtzeiteinsatz auf mobilen Geräten mit beschränkten Ressourcen geeignet ist.

Die ortsbasierten Konsistenzprüfungen und Mechanismen zur Angriffserkennung werden in Bezug auf ihre Genauigkeit und Anwendbarkeit evaluiert. Zu diesem Zweck werden Simulationen für verschiedene MANET-Szenarios mit der Simulationsumgebung JIST/MobNet durchgeführt. Dabei werden insbesondere Angriffe auf die Kommunikationsinfrastruktur betrachtet, bei denen Angreifer das Routingverhalten beeinflussen, um eine falsche Netzwerktopologie vorzutäuschen. Die Evaluierungsergebnisse der Angriffserkennungsmechanismen zeigen für alle Szenarios eine deutliche Differenz der Analysewerte für den Angriffsfall im Vergleich zum Normalbetrieb.

Die erfolgreiche Validierung des Proof of Concepts zeigt, dass die Robustheit der Situation Awareness in taktischen MANETs durch Cross-Data-Analyse erhöht werden kann. Querbeziehungen zwischen Datenquellen lassen sich ausnutzen, um Widersprüche in Datensätzen wirksam zu erkennen. Auf diese Weise wird die Erkennungsleistung verbessert und die Erkennungsrate erhöht. In dynamischen MANET-Umgebungen kann man so falsche Daten durch Sensorstörungen oder Angriffe von bösartigen Knoten besser erkennen. Taktische Teams erhalten so ein präziseres Bild der aktuellen Situation, um ihre Mission erfolgreich umzusetzen.

(2) Wie kann die *Vertrauenswürdigkeit* anderer Netzwerkknoten *effizient und kooperativ* beurteilt werden? Im zweiten Forschungsgebiet liegt der Schwerpunkt auf der Vertrauenswürdigkeit anderer Netzknoten, die Informationen zum Situation-Awareness-Prozess beisteuern. Zu diesem Zweck verbinden wir einen effizienten Broadcast-Gossiping-Ansatz für die Verteilung der Vertrauenswürdigkeitswerte mit der Modellierung von Vertrauenswürdigkeit als (trust,confidence)-Wertepaare. Wir entwickeln effiziente und robuste Mechanismen, die eine kooperative Schätzung der Vertrauenswürdigkeit von Netzknoten im Falle von Angriffen oder Störungen ermöglichen. Die Ergebnisse einer lokalen Beurteilung der Vertrauenswürdigkeit direkter Nachbarknoten (z. B. basierend auf der Erkennung von Routing-Anomalien) werden mit anderen Knoten ausgetauscht.

Die vorgestellte Gossiping-basierte Architektur für die Vertrauenswürdigkeitsschätzung nutzt die besonderen Eigenschaften der drahtlosen Kommunikation in MANETs für die gleichzeitige Verteilung von Vertrauenswürdigkeitsschätzungen an alle direkten Nachbarknoten. Ein effizienter und einfacher Mechanismus zur lokalen Aggregation von Vertrauenswürdigkeitsinformationen begrenzt die Datenmenge, die in das Netzwerk geflutet werden muss. Dies führt zu einer deutlichen Reduktion des Kommunikations- und Verarbeitungsaufwands.

Die Wirksamkeit der Mechanismen zur Vertrauenswürdigkeitsschätzung wird dadurch erhöht, dass bei der Verarbeitung von weitergeleiteten Vertrauenswürdigkeitsinformationen („aus zweiter Hand") auch die Vertrauenswürdigkeit des übermittelnden Knotens betrachtet wird. Für diesen Zweck wird der Vertrauenswürdigkeitswert um einen zweiten Wert (*confidence*) ergänzt, der die Gewissheit der Vertrauenswürdigkeitsbeurteilung bewertet. Diese Modellierung als (trust,confidence)-Wertepaare ermöglicht weiterhin die Abstufung von gefälschten Bewertungen durch das Absenken des zugehörigen confidence-Werts bei widersprüchlichen Einschätzungen.

Wir beschreiben und spezifizieren die Verfahren und Protokolle für eine konkrete Umsetzung der vorgeschlagenen Ansätze im Detail. Zu diesem TEREC-System gehören sowohl Mechanismen für die lokale Beurteilung der Vertrauenswürdigkeit als auch der Gossiping-basierte Austausch von Vertrauenswürdigkeitsinformationen mit direkten Nachbar-Knoten. Wir spezifizieren das Berechnungsmodell für indirekte Einschätzungen der Vertrauenswürdigkeit und für die Kombination von mehreren Vertrauenswürdigkeitsbeurteilungen sowie ein Protokoll für den Gesamtprozess.

Das implementierte TEREC-Modell wird in unterschiedlichen Szenarios durch Simulationen evaluiert. Die Simulationsumgebung JIST/MobNet wird verwendet, um die zeitliche Dynamik des Modells in einer MANET-Umgebung zu analysieren. Weiterhin wird die Leistungsfähigkeit des Konzepts für eine wachsende Anzahl von bösartigen Knoten in einem Grid-basierten Mesh-Netzwerk analysiert. Die Ergebnisse der Evaluation zeigen, dass das System sehr robust ist und dass Vertrauenswürdigkeitseinschätzungen auch in Umgebungen mit mehreren Angreifern schnell zum richtigen Wert konvergieren.

Das entwickelte Konzept für die Vertrauenswürdigkeitsbeurteilung ermöglicht es einer Mehrheit von gutartigen Knoten sich mit ihren Messungen und Einschätzungen durchzusetzen, um so die anderen Netzwerkknoten korrekt zu beurteilen. Die kooperativen Mechanismen zur Vertrauenswürdigkeitsbeurteilung erhöhen die Widerstandsfähigkeit der Situation Awareness gegen Angreifer. Netzwerkknoten, bei denen falsches oder böswilliges Verhalten erkannt wird, werden aus dem Situation-Awareness-Prozess ausgeschlossen. So kann auch im Falle eines Angriffs der Erfolg einer taktischen Mission gewährleistet werden.

(3) Wie können *Systemzustände* auf eine *allgemein anwendbare Weise modelliert* und *umfassend ausgewertet* werden, um gegen falsche und fehlerhafte Daten resistent zu sein? Wir entwickeln ein Konzept für die probabilistische Zustandsmodellierung und -schätzung für Situation Awareness in taktischen MANETs, um die Robustheit gegen ungenaue und fehleranfällige Eingabedaten zu erhöhen. Zustandsschätzungen sind nicht auf einen bestimmten Wert beschränkt sind, sondern werden durch eine „Wahrscheinlichkeitswolke" repräsentiert. Das Konzept basiert auf Partikel-Filter (auch sequentielle Monte-Carlo-Methode genannt) und ist offen und für ein breites Spektrum von Anwendungsszenarios. Dies schließt auch nichtlinearer Zustandsübergängen sowie nicht-gaußschem Rauschen ein. Das Konzept kann auch für hybride Zustandsschätzung angewendet werden bei denen der Systemzustand sowohl diskrete also auch kontinuierliche Zustandsvariablen umfasst. In diesem Fall können die diskreten Variablen z. B. einen Betriebsmodus charakterisieren, der die Modellierung der Systemdynamik basierend auf diskreten Verhaltensmodi ermöglicht.

Der Partikel-Filter-basierte Ansatz ermöglicht die Integration zusätzlicher Informationen an mehreren Stellen. Mehreren Datenquellen sowie zusätzliche Informationsquellen, z. B. Missionsdaten, fachspezifische Vorkenntnisse oder allgemeine Informationsquellen, können in die Zustandsschätzung eingebunden werden. Zusätzliche Informationsquellen können genutzt werden, um Randbedingungen für das Zustandsschätzverfahren vorzugeben, um die Wahrscheinlichkeitsschätzung bestimmter Systemzustände anzupassen oder um unmögliche Zustände auszuschließen.

Wir präsentieren eine Architektur für die verteilte Zustandsschätzung, die die besonderen Herausforderungen in verteilten taktischen MANETs adressiert. Für die verteilte, kooperative Zustandsschätzung wählt jeder Knoten eine geeignete Teilmenge der lokale verfügbaren Messungen und Zustandsschätzungen aus und verbreitet diese innerhalb des taktischen MANETs.

Beim Austausch von Messungen und Zustandsschätzungen innerhalb des MANETs ist zu beachten, dass Informationen, die von anderen Netzwerkknoten empfangen werden, sich auf einen Zeitpunkt in der Vergangenheit beziehen können. Für diesen Fall wird ein Re-Simulation-Mechanismus vorgeschlagen. Dieser dient zur Auflösung der zeitlichen und kausalen Zusammenhänge von Messungen und Zustandsschätzungen, die von anderen Netzknoten empfangen werden.

Die Konzepte werden auf ein typisches Anwendungsszenario für taktische Situation Awareness angewendet: Task Force Tracking (TFT). Der vorgestellte TFT-Mechanismus betrachtet zusätzliche Informationsquellen wie missionsspezifische Informationen (z. B. Mobilitätsmodelle) und topographische Daten. Das TFT-Modell wird implementiert und in einer Simulationsumgebung für verschiedene MANET-Szenarios mit unterschiedlichen Mobilitäts- und Funkausbreitungsmodellen evaluiert. Die Ergebnisse zeigen, dass TFT durch die Zustandsschätzung mit Partikel-Filtern unter der Einbeziehung von externen Informationsquellen sowie der Netzwerkfähigkeiten von taktischen MANETs signifikant verbessert werden kann. Die Simulationsergebnisse zeigen für verschiedene Szenarios mit unterschiedlichen Mobilitäts- und Funkausbreitungscharakteristiken, dass das vorgeschlagene TFT-Modell sowohl die Genauigkeit wie auch die Robustheit der Positionsschätzungen im Vergleich zu Standard-TFT-Modellen verbessert.

Das vorgestellte Konzept erhöht die Robustheit der Zustandsschätzung in Bezug auf fehlerhafte Datenquellen. Die Partikel-Filter-basierte Zustandsschätzung kann grundsätzlich flexibel für jede Art von relevanten Situationsinformationen angewendet werden. Die Einbeziehung zusätzlicher Informationsquellen verbessert die Genauigkeit der Zustandsschätzergebnisse und macht das Situation-Awareness-Bild präziser. Auf diese Weise können taktische Einsatzkräfte besser unterstützt werden, um ihre Mission selbst im Falle von Angriffen oder Systemfehlern erfolgreich abzuschließen.

Die drei in dieser Dissertation vorgestellten Konzepte erhöhen die Robustheit von Situation Awareness in taktischen MANETs. Die *Cross-Data-Analyse* hilft in dynamischen MANET-Umgebungen, um gegen falsche Daten aufgrund von Sensorfehlern oder bösartigem Verhalten von angreifenden Knoten vorzugehen. Widersprüche in Datensätzen lassen sich durch die Nutzung von Querbeziehungen zwischen Datenquellen effektiv erkennen. *Kooperative Mechanismen zur Beurteilung der Vertrauenswürdigkeit* führen zu einer Verbesserung der Robustheit des Situation-Awareness-Prozesses gegen Angreifer. knoten, die für falsche oder böswillige Verhalten verantwortlich sind, können identifiziert und vom Situation-Awareness-Prozess ausgeschlossen werden. Die *probabilistische Zustandsmodellierung* erhöht die Robustheit des Verfahrens in Bezug auf fehlerbehaftete Messungen. Zustandsschätzungsergebnisse sind genauer und das Situation-Awareness-Bild ist realistischer. Auf diese Weise erhalten taktische Teams ein präziseres Bild der aktuellen Situation. Einsatzkräfte können so gut wie möglich unterstützt werden, um auch im Falle eines Angriffs oder bei Systemfehlern den erfolgreichen Abschluss ihre Mission zu gewährleisten.

Zukünftige Forschungsarbeiten Zum Abschluss wird einen Analyse von möglichen zukünftigen Forschungsthemen sowie potenziellen Erweiterungen der präsentieren Konzepte präsentiert. Zunächst sollte untersucht werden, wie sich die entwickelten Konzepte in den gesamten Situation-Awareness-Zyklus integrieren lassen, einschließlich Situation Projection (Ebene 3), Entscheidungsfindung und Umsetzung von Maßnahmen. Zweitens könnte die Interaktion und Integration der drei großen Bereiche dieser Dissertation weiter ausgearbeitet werden. Die Konzepte zur Cross-Data-Analyse und zur probabilistische Zustandsmodellierung wurden bereits kombiniert indem Missionsinformationen und zusätzliche Wissensquellen bei der probabilistische Zustandsschätzung verwendet werden.

In taktischen Szenarios ist durchaus möglich, dass die ad-hoc-basierte Kommunikationsinfrastruktur durch die (zeitweise) Anbindung an eine leistungsfähige Back-End-Infrastruktur ergänzt wird. Daher könnte untersucht werden, wie die entwickelten Konzepte für Hybrid-Szenarios angewendet und ergänzt werden könnten. In taktischen MANETs sind oft verschiedene Organisationen, wie z. B. THW und Rotes Kreuz, parallel aktiv. Daher sollte analysiert werden, wie die entwickelten Konzepte in Bezug auf die spezifischen Anforderungen und Herausforderungen in Koalitionsumgebungen optimiert werden könnten.

Schließlich könnte das vorgeschlagene Konzept zur probabilistischen Zustandsmodellierung für die hybride Zustandsschätzung weiterentwickelt werden. Die einzelnen Schritte des auf Partikel-Filter basierenden Zustandsschätzungsprozess sollten in Bezug auf mögliche Erweiterungen sowie Optimierungen für diskrete Zustände untersucht werden.

Contents

I. Motivation and Background 1

1. Introduction 3

1.1. Motivation . . . 4

1.1.1. Mobile Ad Hoc Networks . . . 4

1.1.2. Tactical Environments . . . 5

1.1.3. Situation Awareness in Tactical Mobile Ad Hoc Networks . . . 7

1.2. Research Challenges . . . 10

1.3. Structure of the Dissertation . . . 12

1.4. Conclusions . . . 14

2. Problem Definition 17

2.1. Situation Awareness in Tactical MANETs . . . 17

2.1.1. Situation Awareness . . . 20

2.2. Technical Setup . . . 23

2.2.1. Operational Environment . . . 23

2.2.2. Example Mission Scenario . . . 26

2.2.3. Specific Characteristics of Tactical MANETs . . . 27

2.2.4. Adversary Model . . . 28

2.3. Objectives . . . 30

2.3.1. Situation Awareness in Tactical MANETs . . . 30

2.3.2. Design Objectives . . . 31

2.3.3. Research Challenges . . . 32

3. Related Work 37

3.1. Situation Awareness 37
3.1.1. Situation Awareness Definition 38
3.1.2. Errors and Cyber Threats in Situation Awareness 42
3.1.3. Situation Awareness in Tactical MANETs 46
3.1.4. Detection of Malicious Behavior in MANETs 48
3.2. Cross Data Analysis 54
3.2.1. Data Fusion 54
3.2.2. Plausibility and Consistency Checks 55
3.2.3. Cross-Layer Approaches for Wireless Mobile Networks 60
3.2.4. Cross Data Analysis of Location Related Data 64
3.3. Cooperative Trust Assessment 78
3.3.1. Trust Modeling 78
3.3.2. Trust Assessment 81
3.3.3. Trust and Cooperation 84
3.4. Probabilistic State Modeling 92
3.4.1. State Modeling and Estimation 92
3.4.2. Particle Fiter 96
3.4.3. State Estimation Approaches based on Particle Filter 104

II. Research on Robust Situation Awareness 113

4. Cross Data Analysis 115
4.1. Data Sources 117
4.1.1. Direct Sensor Data 117
4.1.2. Indirect Sensor Data 120
4.1.3. Mission-Specific Knowledge Sources 122
4.1.4. General Knowledge Sources 123
4.2. Cross-Relationships and Consistency Checks 123
4.2.1. Overlaps of Data Sources 123
4.2.2. Cross-Relationships 124
4.2.3. Consistency Checks 125
4.3. Cross Data Analysis of Location Related Data Sources 130
4.3.1. Location Related Data Sources 130

4.3.2. Topography Based Radio Propagation Model 135
4.3.3. Cross-Relationships Between Data Sources 141
4.3.4. Consistency Checks and Attack Detection 144
4.4. Summary 149

5. Cooperative Trust Assessment 151
5.1. Trust Assessment in Tactical MANETs 151
5.1.1. Specific Properties and Challenges 151
5.1.2. Design Objectives 153
5.2. Efficient Cooperative Trust Assessment 154
5.2.1. Overview 155
5.2.2. Flooding based Trust Assessment Architecture 157
5.2.3. Gossiping based Trust Assessment Architecture 158
5.2.4. Performance Evaluation and Comparison 162
5.3. Enhanced Gossiping based Trust Assessment Architecture 168
5.3.1. General Idea 168
5.3.2. Weighted Combination of Trust Values 169
5.3.3. Enhanced Trust Assessment Model 170
5.4. TEREC 171
5.4.1. Overview 172
5.4.2. Trust Assessment and Modeling 173
5.4.3. Gossiping based Trust Assessment Architecture 175
5.4.4. Trust Value Combination 176
5.4.5. Trust Assessment Protocol 181
5.5. Summary 182

6. Probabilistic State Modeling 185
6.1. Probabilistic State Modeling and Estimation 185
6.1.1. State Modeling 185
6.1.2. Probabilistic State Estimation 189
6.1.3. State Estimation based on Particle Filters 192
6.2. Distributed Particle Filter based Approach 195
6.2.1. Distributed State Estimation Architecture 195
6.2.2. Data Selection and Exchange 197

6.2.3. System Update 199
6.2.4. Situation Assessment 201
6.2.5. Distributed Particle Filter Protocol 202
6.3. Probabilistic State Modeling for Task Force Tracking 205
6.3.1. Task Force Tracking 205
6.3.2. Distributed TFT based on PF 206
6.3.3. SMC-TFT Protocol 209
6.4. Summary 210

III. Evaluation of Research Results 213

7. Simulation and Evaluation 215
7.1. Simulation Environments 215
7.1.1. ns-2 217
7.1.2. JiST/MobNet 219
7.1.3. PF Network Simulator 221
7.2. Cross Data Analysis: Location Related Information 222
7.2.1. Distance Estimation based on Radio Signal Characteristics . . . 223
7.2.2. Topography based Radio Propagation Model 232
7.2.3. Location based Consistency Checks and Analysis 241
7.3. Cooperative Trust Assessment: TEREC 246
7.3.1. Mobile Ad hoc Environment 248
7.3.2. Mesh Network 249
7.3.3. Quantitative Evaluation of the Trust System 254
7.4. Probabilistic State Modeling: Task Force Tracking 255
7.4.1. Simulation Setup 256
7.4.2. Evaluation Results 258
7.5. Summary 263

8. Summary and Outlook 267
8.1. Summary of Contributions 269
8.2. Future Research 274

Appendices 279

A. Publications and Talks 279

A.1. Publications 279
A.2. Talks 280
A.3. Other Publications 282

B. Supervising Activities 285

B.1. Diploma and Master Theses 285
B.2. Bachelor Theses 286
B.3. Other Supervised Theses 287

C. Curriculum Vitae 289

Bibliography 291

List of Tables

3.1. Definitions of Relevant Symbols for State Estimation 93
3.2. Filtering via Sequential Importance Sampling [RAG04] 102
3.3. Resampling Algorithm [RAG04] 103

4.1. Consistency Checks based on Location Related Data Sources 146

5.1. Evaluation of Trust Assessment Architectures 167

6.1. Notation for Probabilistic State Modeling and Estimation 186
6.2. Examples of State Information $\mathbf{x}_k^i$ in Tactical MANETs 187

7.1. Experimental Setups 224
7.2. Equipment Used for the Experiments 224
7.3. Measurement Results of Application Evaluation 231
7.4. Recommended Parameter Settings for Reflection and Diffraction Depth 234
7.5. Computation Times 236
7.6. Experimental Setups with Obstacle 237
7.7. Simulation Parameters for Different Evaluation Setups 241
7.8. Number of Attackers for Different Simulation Setups 242
7.9. Time Until Attacker is Detected by a Certain Percentage of Nodes . . . 250
7.10. Number of Rounds Until Attacker is Detected by a Certain Percentage of Nodes 255
7.11. Number of Messages and Average Estimation Error vs. Message Generation Interval 262

List of Figures

1.1. Levels of Response: Strategic, Operational and Tactical 6
1.2. Situation Awareness (SA) 7

3.1. Situation Awareness Model based on Endsley [End95, BP02] 39
3.2. Team Situation Awareness Model [SSW*08, SSWJ09] 42
3.3. Distributed Situation Awareness Example [SSW*08] 43
3.4. Architecture of a Cooperative MANET IDS [ZLH03, ZL04]. 50
3.5. JDL Model of Data Fusion [HL01, HM04] 55
3.6. Levels and Aspects of Knowledge Inconsistency [Ngu08]. 59
3.7. Averaging and Observation Windows (based on [ZMHT05]) 84
3.8. Trust Propagation in Parallel Network via P Paths (based on [MMH02]) 88
3.9. Hybrid State Space Model [DLNP05, Kaw12] 94
3.10. Hybrid State Robot Example [Fun04] 106
3.11. Example Scenario of Out-of-Sequence Measuremens (OOSM) and In-Sequence-Measurements (ISM) (based on [ZBS11]) 110

4.1. Generic Model of Cross Data Analysis: Different Data Sources Providing Various Data Sets, Contained Cross-Relationships and Targeted Entities of the Assessment 116
4.2. Generic Sensor Model 118
4.3. Extraction and Utilization of Indirect Sensor Data 121
4.4. Overlaps of Data Sources in Respect to Time, Space and Subject 124
4.5. Wormhole Attack by Attacker X and X' Acting as Legitimate Network Node C 128
4.6. Absolute Geo Positions and Distances Between MANET Nodes 133
4.7. Reflection on a Flat Surface 137

4.8. Geometry and Components of the Deflected Field 139
4.9. Cross-Dependencies of Geo Coordinates, Routing Information and Radio Signal Strengths 142
4.10. Distance Verification of a Four Hop Route from Node *A* to Node *B* . . . 146
4.11. Radio Signal Strength Checking 148

5.1. Local Trust Assessment 174
5.2. Gossiping based Cooperative Trust Assessment 176
5.3. Combination of Local Trust Assessments and Neighbor Trust Estimations for Combined Trust Estimation 176
5.4. Trust Calculation for Distant Nodes 177
5.5. Calculation of Trust Value for Distant Nodes Example 178
5.6. Trust Calculation based on Multiple Estimations 180
5.7. Chain of Calculations in order to Estimate the Trustworthiness of a Node 181

6.1. Example for Continuous State Modeling: Location Tracking 187
6.2. Example for Discrete State Modeling: OLSR Link State 188
6.3. Location Tracking Example: Incorporation of Constraints 191
6.4. Data Fusion Concept 196
6.5. Distributed Particel Filter Concept 203
6.6. Basic Concept of SMC-TFT 209

7.1. Processing Chain of JiST (based on [Bar04c, EHKP06, EHK*07]) . . . 220
7.2. Component-based Architecture of MobNet (based on [KBHS07,EHK*07])221
7.3. Experiment 1: Different Antenna Orientations 225
7.4. Notebook Orientations in Relation to Each Other 226
7.5. Experiment 2: Different Surfaces 227
7.6. Experiment 3: Different WLAN Cards 227
7.7. Screenshot of the Application 230
7.8. Distances Estimated Using Application vs. Real Distances 230
7.9. iNSpect Validation Tab 233
7.10. Connectivity Between Nodes for Example Scenario. 236
7.11. Evaluation Setups with Obstacle: a) Experiment 4 (Reflection) and b) Experiment 5 (Diffraction). 237

7.12. Experiment 4: Scenario with Obstacle (Reflection) 238
7.13. Experiment 4: Scenario with Obstacle (Diffraction) 239
7.14. Relative Frequency of Warnings per Node and Analysis Process of the *Distance Verification* in Different Scenarios 244
7.15. Relative Frequency of Warnings per Node and Analysis Process of the *Connectivity Examination* in Different Scenarios 246
7.16. Trust Values provided by Neighbours that are Considered on Updates . 247
7.17. Percentage of Detected a Simulated Attacker with AODV 249
7.18. Progress of Detecting a Simulated Attacker in an AODV based MANET 250
7.19. Environment with Two Malicious Nodes 251
7.20. Trust Progress of Node N0, N13, N22 about Node N28 with Two Malicious Nodes 252
7.21. Environment with Eleven Malicious Nodes 253
7.22. Trust Progress of Node N0, N13, N22 about Node N28 with Eleven Malicious Nodes 254
7.23. Average Estimation Error vs. Node Density for Manhattan Grid Model . 259
7.24. Average Estimation Error vs. Node Speed for Manhattan Grid Model . . 260
7.25. Average Estimation Error vs. Group Speed for RPGM Model 260
7.26. Radio Propagation Model and Radio Range 261
7.27. GPS Mesurement Accuracy 262

Part I.

Motivation and Background

1. Introduction

Natural catastrophes (e. g., earthquakes, tsunamis or hurricanes) or human-induced disasters (e. g., chemical plant explosions or terrorist attacks) often have significant impacts on the population and the infrastructure in the affected area. The objective of rescue and disaster recovery forces[1] is to provide technical and humanitarian relief in such situations [Byt05]. Task forces take care of victims and provide required supplies (e. g., drinking, food or tents). Rescue and disaster recovery forces typically set up temporary infrastructure while existing infrastructure is rebuilt and put back into operation. In order to minimize the effects of such catastrophes rescue forces need to assess the situation after the disaster, including extent of damage and numbers and types of casualties. For example following the Haiti earthquake in 2010 [RL10] rescue forces provided technical and humanitarian aid to affected areas. Small tactical teams searched for victims in difficult to access areas on the island. Their mission was particularly challenging as the existing infrastructure was broken. They needed to be aware of the current situation in a dynamically developing scenario in order to react quickly to save lives.

Appropriate, efficient and effective IT (information technology) support is essential for the success of a mission. Task forces need to be provided with adequate mobile applications for team and resource management, victim support and situation assessment, e. g., location of victims, team situation, presence of harmful substances or buildings in danger of collapse. These tools are the basis for situation awareness enabling target-oriented command & control in order to achieve mission objectives. Reliable and accurate state estimation is the basis technology for effective situation awareness applications. The primary objective explored in this dissertation is increasing the robustness of state estimation. In the described scenarios the communication infrastructure may

[1] e. g., civil protection organization as the Bundesanstalt Technisches Hilfswerk (Federal Agency for Technical Relief, THW)

be destroyed (or should not be used for other reasons) and a backup communication infrastructure needs to be set up. Mobile devices connected by a mobile ad hoc network (MANETs) are a quick, flexible and efficient way to provide such an infrastructure.

1.1. Motivation

Technological progress in IT enables the creation of new tools and techniques for mission support in disaster recovery scenarios and tactical environments. Mobile devices have rapidly evolved in the last ten years providing computing and networking capabilities that are on the same level as former workstations. Substantial amounts of data may be stored on mobile devices providing additional information sources (e. g., maps, technical descriptions) and some new sensors delivering dynamic data are becoming commonly available (e. g., GPS receiver, accelerometer, camera). This enables new possibilities for services and applications which should be exploited to provide tailored tools for tactical networks.

In parallel to these opportunities there are new threats and challenges that arise for IT systems due to technological progress and other general developments and changes. Cyber attacks become easier to implement and feasible for the general public as tools are freely available online or provided as commodity by cybercrime businesses. Terrorists may even attack rescue forces and IT system after the initial attack was launched in order to increase the extent of the damage caused and the number of casualties. The target of their actions is to disturb regular operations in as many ways as possible which leads to a strong demand for resilience and availability of back-up infrastructures.

This dissertation explores the implications of these new capabilities and threats as they relate to tactical MANET environments.

1.1.1. Mobile Ad Hoc Networks

Mobile ad hoc networks (MANETs) are utilized in tactical environments as they provide a quick and flexible way to implement a backup communication infrastructure when the normal infrastructure is destroyed or should not be used for other reasons.

The main characteristics of MANETs (as we comprehend them within this context) are infrastructureless operation, node mobility and ad hoc wireless communication. Within a decentralized network, all nodes are generally sender, receiver and router at the same time. The network is formed by the participating nodes involving without relying on (fixed) infrastructure (such as base stations of access points) or other central instances, e. g., for storing routing information. Nodes do not have prior knowledge of the network topology, the network is based on self-configuration and self-maintenance.

Due to node mobility the network topology may dynamically change in MANETs at any time. The resources of the mobile nodes are constrained in respect to energy as well as processing power, main memory and storage capabilities. In general, a communication link to a back-end infrastructure and/or Internet is not present but may be temporarily available. Communication is based on a wireless broadcast medium with changing link quality, e. g., due to interference. The bandwidth and transmission range are limited and multi-hop routing is require in order to reach nodes that are further away.

MANETs are generally vulnerable due to the fact that wireless communication can be eavesdropped and disturbed remotely and that node mobility causes highly dynamic network behavior, e. g., frequent route finding and updating, and imposes strict resource constraints on the network nodes. Other threats such as node compromise and Byzantine failures, fraudulent route propagation, and denial of service attacks are more specific to tactical environments. Providing robustness and detection capabilities against such attacks will benefit the general MANET case as well as concepts may be adapted for other application areas.

Research on ad hoc networks, particularly MANETs, is anticipated to have significant applications in environments in which fixed infrastructures may not be feasible or ineffective.

1.1.2. Tactical Environments

The application of MANETs in tactical networks is particularly promising but even more challenging. In *tactical networks* (as opposed to strategic networks) the focus is on the achievement of immediate tactical objectives as illustrated in figure 1.1 (based on [Cou05]). Tactical MANETs are typically deployed by first responders in case of

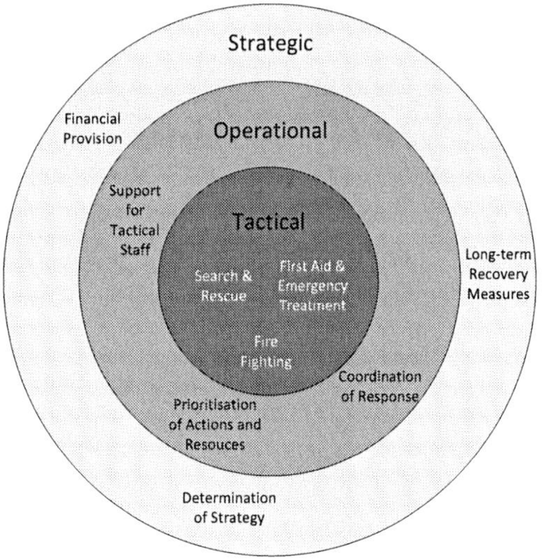

Figure 1.1.: Levels of Response: Strategic, Operational and Tactical

disasters, e. g., hurricane Katrina [Bla09]. Even in developed countries communication networks may not be available in such scenarios. According to [War05] emergency communication networks in most cities of the United States would not survive major natural disasters or terrorist attacks.

There are several constraints and characteristics in tactical MANETs that do not typically hold for MANETs in general. There is typically a closed group of participants that are previously known and cooperate sharing a common goal. In contrast to other application scenarios for MANETs the devices are not arbitrarily mixed devices but standardized equipment is used according to well-defined specifications for specific purposes and mission scenarios. Multiple groups and organizations sometimes need to join their resources to communicate and collaborate for a specific task or mission. The policy is not to tolerate and mitigate malicious behavior but to detect and to terminate it. These properties need to be taken into account when providing situation awareness in tactical MANETs.

1.1.3. Situation Awareness in Tactical Mobile Ad Hoc Networks

Situation Awareness The assessment and the awareness of the current situation of all relevant elements in the environment are crucial in tactical MANETs to protect the action force and to support the mission objectives on site. According to Endsley [End95, EC08] *situation awareness* is "the perception of elements in the environment within a volume of time and space, the comprehension of their meaning, and the projection of their status in the near future" (as shown in figure 1.2 [EC08]).

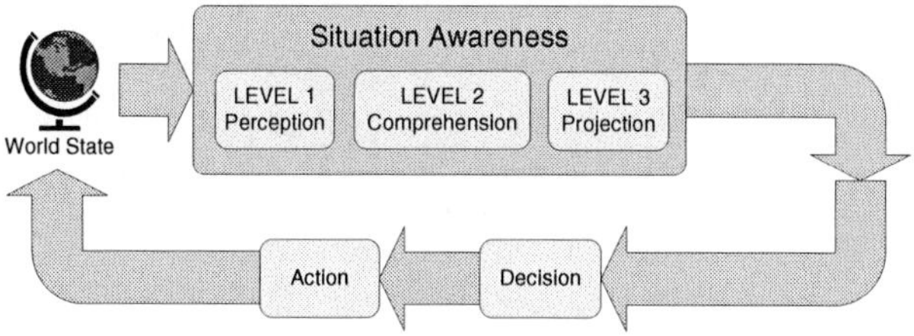

Figure 1.2.: Situation Awareness (SA)

This means that all information that is available should be used to create an awareness of what is happening around a particular team member and to project the impact of the current situation together with planned actions and mission objectives on the near future. The goal is to enable conscious and informed decisions by the user which trigger respective actions which then change the world state. This feeds back into the situation awareness process in order to check if the expected impact was achieved.

The focus of this dissertation is on state estimation, represented by perception (level 1) and comprehension (level 2), which is the basis for the other steps. State estimation forms the core part which is mostly related and correlated to the specific properties of tactical MANETs. For the prediction step (level 3), user interaction and decision support specific solutions based on application scenario and domain knowledge need to be developed. In general, existing concept may be adopted for the other steps, which should be further elaborated in future research.

The main area of interest for perception and comprehension are entities composing the actual tactical MANET: task force members and their equipment, e. g., communication devices and vehicles. These components represent the particular challenges, peculiarities and opportunities for situation awareness in tactical MANETs. Tracking and management of other resources and external sensor data are of secondary importance within this dissertation as they are common problems which are addressed in other fields as well. However, additional entities and data sources related to the operational environment are considered whenever they provide some additional information and value for state estimation and situation awareness of the core components.

Within the scope of tactical MANETs the core challenge is to assure security and safety of the operational team and the successful completion of overall mission objectives. Communication disruptions, failures or attacks may lead to missing or false information. As a result, delayed reaction or wrong decisions may crucially affect the operational team and the mission objectives, potentially even causing casualties. Therefore the detection of specific threats and attacks, e. g., compromised equipment, is a core part of the situation awareness process.

Situation Awareness in Tactical Mobile Ad Hoc Networks In the following we present an overview of solutions that were proposed in related work for situation awareness and state estimation in tactical networks. The presentation is grouped based on the technical aspects that are specifically addressed within the proposed solutions.

A major part of related work focues on the *underlying communication protocols* in tactical environments. For example, the OASIS [Cou05, Cou08] and DistressNet [GZC*10] projects provide a communication infrastructure based on wireless ad hoc and networks for supporting firefighters, police and medical staff in disaster response. Most of these solutions also address the *localization* of task force members. For example, DistressNet provides accurate resource localization for distributed collaborative sensing to provide first responders with detailed situational awareness.

Another important aspect is *information management* as situation awareness relies on exchanging and compiling collected information on a structured basis. A standard for the description of tactical situations, the so called Tactical Situation Object (TSO) was defined within the OASIS [Cou08] project. The SoKNOS (Service-Oriented ArchiteCtures Supporting Networks of Public Security) project [PPBS10, PDTS*09] proposes service-oriented solutions enabling ad-hoc information sharing and the flexible integration of heterogeneous information sources in emergency scenarios. As units move and the situation dynamically changes the relevance of information changes as well.

Other work focuses on *visualization, human-computer interaction and user experience*. They try to optimize and evaluate the utility, ease of use and efficiency of the tactical situation awareness systems. A Common Operating Picture (COP), displaying messages and resources in a geospatial map-centric system, can be used to create situation awareness.

There are some efforts focusing *system monitoring for security incidents* in tactical MANETs. The detection of misbehaving nodes was also considered as one of the requirements for tactical situation awareness within the OASIS project [Cou08]. However, they left this as an open issue as they claimed that no suitable misbehavior detection solutions were available yet. MANET IDS architectures proposed for other applications scenarios may be adapted and applied within tactical environments as well (see [XPS11] for an overview). These efforts are, however, looking at attackers and security incidents only and are limited on security related aspects; they do not address situation awareness in general.

Shortcomings of Existing Work The related work presented above provides some promising solutions and building blocks for situation awareness in tactical MANETs. However, it either provides only *partial solutions* for specific areas or *neglects some important aspects required for robustness*. Many efforts focus on specific sub-problems (such as MANET communication, localization, human-computer interaction, information management or security monitoring) and do not address the overall problem.

There are some efforts focusing on security monitoring (IDS) in tactical MANETs, which, however, do not address situation awareness in general. Besides these efforts potential *security risks and threats to situation awareness and state estimation*, e. g., node compromise or injection of false data by malicious nodes, are typically neglected or only marginally addressed. The proposed situation awareness systems do not provide protection against malicious nodes or counter-measures against false or malicious data.

The proposed solutions do not comprehensively address the challenges resulting from the *distributed and cooperative nature of state estimation and situation awareness in tactical MANETs*. Relevant information needs to be collected, exchanged, fused and assessed in a collaborative fashion. Particularly the trustworthiness of (sensor) data provided by remote nodes is not sufficiently considered and incorporated within the situation assessment processes.

The proposed state estimation solutions are *not robust enough, they do no directly consider false (faulty and error-prone) or malicious data.* (Sensor) data is aggregated and fused in order to estimate a value for each system parameter and derive a specific situation picture. No special means are incorporated for dealing with inconsistent data in order to deal with faulty sensors or attackers. The available (potentially overlapping and correlated) *data sources are not comprehensively utilized and exploited.* In particular cross-relationships between data sets are not used to detect errors or inconsistencies.

1.2. Research Challenges

The goal of this dissertation is to address most prominent shortcomings of existing solutions. Mobility and related properties are one of the major challenges in MANET environments. Therefore location-related aspects are particularly addressed within this dissertation. The main objective is therefore increasing robustness in state estimation for situation awareness in tactical MANETs. This objective is broken down into three research challenges represented by the following three questions:

1. How can location-related *data sources* be utilized in a *comprehensive and effective way* to detect malicious behavior?
2. How can the *trustworthiness* of other nodes be *efficiently and cooperatively* assessed?
3. How can *system states* be *modeled in a generally applicable way* and *comprehensively assessed* to counter inaccurate and error-prone data?

In the following we describe the three research challenge addressed within this dissertation (see also section 2.3.3 on page 32):

Utilizing all available location-related data sources in a comprehensive and effective way to detect malicious behavior All available information should be exploited (in particular including extrinsic information sources in respect to network communication) in order to provide a comprehensive awareness of the current situation to the users. Present approaches for the assessment of the security situation in tactical MANETs address mainly the core communication protocol and system. They try to detect attacks based on routing protocol data and neglect application layer information and other information sources that may be available without any direct usage or benefit. The challenge is to use the broad set of data sources available in tactical MANETs in a comprehensive and effective way. Data may be provided by dynamic direct or indirect sensor data and static mission-specific or general knowledge sources (e. g., mission profiles, mission profile, tactical mobility patterns, topography model). This includes local data sources as well as data provided by other nodes. There is some redundancy within data sources as data sets may partially overlap or have some cross-relationships. It needs to be investigated and modeled how this can be exploited for detecting false or malicious data in order to increase robustness of state estimation for situation awareness.

Efficient cooperative assessment of the trustworthiness of other nodes Trust is the basis for robust cooperation in order to obtain a common situational picture. In tactical environments all participants share a common goal and mission and there is a sharp borderline between benign nodes (that are a legitimate part of the network) and malicious nodes (attackers). Therefore, specifically tailored cooperation mechanisms are required for distributed trust assessment in tactical MANETs that are particularly robust to attacks or node failure. A robust communication model needs to be developed which works well in a wireless ad hoc network with a lossy transmission channel and dynamic network topology changes. Each node should be judged independently by combining local misbehavior detection with information provided by distant nodes. Therefore, trust modeling should also reflect certainty and accuracy of trust estimations and provide robust mechanisms for incorporating indirect trust assessments.

Modeling of system states in a generally applicable way and comprehensive state assessment to counter inaccurate and error-prone data As current state estimation solutions for situation awareness in tactical MANETs do not provide sufficient robustness in respect to inaccurate and error-prone data, suitable concepts for state modeling and estimation need to be developed in order to solve this problem. The developed mechanisms should be generally applicable and allow to flexible incorporating of multiple observations and additional information sources into the state estimation process exploiting all available data sources including observations and estimations provided by other network nodes. The state estimation architecture needs to meet the challenges resulting from the distributed nature of tactical MANETs. Faulty sensors may lead to false observations and attacker may intentionally inject wrong data into the situation awareness system. Therefore state modeling and estimation needs to provide the means to deal with inconsistend or wrong data in order to increase robustness against failures and false behavior. The developed mechanisms should be open and applicable to a wide range of application scenarios.

The potential of the cooperative nature of MANETs has to be exploited and available data sources comprehensively utilized. The challenges resulting from the distributed nature have to be addressed in order to ensure trustworthiness of other network nodes and mitigate malicious behavior. Resilience mechanisms are required to cope with error-prone measurement data and system faults. Combining the solutions for these three research challenges enables robust situation awareness in tactical MANETS.

1.3. Structure of the Dissertation

In the following we give a short overview on the structure of this dissertation.

Part I (*Motivation and Background*) describes the motivation for this dissertation and provides some background information.

The next chapter (*Problem Definition*) describes the addressed problem in more detail on a technical level. Typical application areas for tactical MANETs in emergency response and disaster recovery are presented and it is discussed what situation awareness means within this context. An overview of the assumptions about the specific characteristics of tactical MANETs is given and an adversary model is specified. Finally, the objectives and research challenges for robust situation awareness in tactical MANETs are described.

In chapter 3 (*Related Work*) we analyze the state of the art regarding situation awareness in tactical MANETs, state estimation and related fields. Related work is presented in respect to the solutions addressing the three research challenges addressed within this dissertation: cross data analysis, cooperative trust assessment and probabilistic state modeling.

In part II (*Research on Robust Situation Awareness*) we further elaborate each of these research challenges and present our solutions.

In chapter 4 (*Cross Data Analysis*) we give an overview of data source available in tactical MANETs, we discuss potential overlaps of data sources and we analyze different kinds of cross-relationships which may exist even in the case of non-overlapping data sets. Based on this discussion we show how plausibility and consistency checks can be exploited in cross data analysis to increase robustness of situation awareness and apply the proposed mechanisms to location related data sources as a proof of concept.

In chapter 5 (*Cooperative Trust Assessment*) we analyze how nodes can cooperate in tactical MANETs in a trusted and robust way and develop cooperative mechanisms that allow the assessment of the trustworthiness of other network nodes, particularly in the case of attacks or node failures. The proposed mechanisms that address the specific requirements in tactical MANETs, particularly efficiency, and show how they can be enhanced to incorporate also the trustworthiness of other network node into indirect trust assessment.

In chapter 6 (*Probabilistic State Modeling*) we present concepts for probabilistic state modeling and estimation and discuss how they can utilized in order to make situation awareness in tactical MANETs more robust in respect to inaccurate and error-prone input data. We propose a concept based on particle filters that is open and generally applicable to a wide range of application scenarios. It enables the incorporation of additional information at multiple stages. A specific challenge is the distributed nature of tactical MANETs which is addressed by a distributed state estimation architecture using re-simulation. Finally, we present Task Force Tracking (TFT) as an example application which shows how these concepts can be implemented for a specific application scenario.

Part III (*Evaluation of Research Results*) presents the evaluation of the research results, summarizes the dissertation and gives an outlook on future research.

In chapter 7 (*Simulation and Evaluation*) we evaluate the concepts that were presented in the three previous chapters. For the implementation and validation we select suitable tools that allow to show the most important aspects in an efficient and effective way. In the following the simulation environments that were chosen for the evaluation and further develop are presented. Then the evaluation scenarios for all three main contribution area (cross data analysis, cooperative trust assessment and probabilistic state modeling) are presented and the evaluation results are discussed in detail. The evaluation results are summarized and some conclusions are drawn.

Finally, in chapter 8 (*Summary and Outlook*) we summarize our main contributions and give an outlook on future research.

1.4. Conclusions

The overall research objective of this dissertation is to improve situation awareness in tactical MANETs for emergency response and disaster recovery. The main focus is on improving state estimation in tactical MANETs as this is the basis for increasing robustness of situation awareness, e. g., in respect to error-prone and/or malicious data. In the following we present the proposed concepts addressing the three identified research challenges and summarize our main contributions.

Firstly, we develop a general concept for *cross data analysis* that enables the incorporation of direct and indirect sensor data and additional knowledge sources (such as mission-specific knowledge and general information sources) in order to detect inconsistencies. Cross data analysis analyzes multiple data sources collectively in order to detect malicious data. Contradictions in data sets can be more effectively detected exploiting cross-relationships between data sources. This improves the detection capabilities and increases detection rates. In dynamic MANET environments this helps to cope with wrong data due to sensor faults as well as with malicious behavior of attacking nodes. This way tactical teams get a clear and correct picture of the situation.

Secondly, we develop efficient and robust mechanisms for *cooperative trust assessment* in tactical MANETs. For these purposes we combine efficient broadcast gossiping trust aggregation and trust information modeling as (trust, confidence) value pairs. When incorporating second hand trust information also the trustworthiness of the providing node is considered. An efficient and simple information local trust aggregation mechanism limits the amount of data that is flooded into the network significantly reducing communication and processing overhead. The proposed TEREC model is implemented and evaluated based on simulations in various scenarios. The trust assessment scheme allows a benign majority of nodes to prevail and accurately classify network nodes based on observations and trust estimations. The cooperative trust assessment mechanisms improve the resistance of the situation awareness to attackers. Network nodes responsible for wrong or malicious behavior can be identified and excluded from the situation awareness process. This can ensure the success of a tactical mission even in the case of attack.

In the third part we develop a concept for *probabilistic state modeling and estimation* for situation awareness in tactical MANETs to increase the overall robustness regarding inaccurate and error-prone input data. The proposed concept based on particle filters is open and applicable to a wide range of application scenarios and allows incorporating of additional information at multiple stages. We present a distributed state estimation architecture where observations and state estimates are selected and exchanged within tactical MANETs for improved cooperative state estimation. The concepts are applied to a specific application scenario: Task Force Tracking, a typical example application for tactical situation awareness. Probabilistic state modeling increases the robustness of the state estimation process to measurement noise. State estimation based on particle filters can flexibly be applied for basically any kind of relevant situation information. Incorporating additional information sources improves the preciseness of state estimation based on error-prone data sources. State estimation results are more accurate and the situation awareness picture is more realistic. This way tactical task forces can be supported as good as possible even in case of attack in order to successfully complete their mission.

2. Problem Definition

In this chapter the problem that is addressed within this dissertation is described in more detail on a technical level. Therefore typical application areas are presented and situation awareness in tactical MANETs is described. Within this dissertation the term tactical MANETs represents the team members of a task force fulfilling a specific mission in emergency response and their equipment, which may also include some vehicles. An overview of the assumptions about the specific characteristics of tactical MANETs is given. Finally, the overall objectives for robust state estimation mechanisms for situation awareness in tactical MANETs are specified including design objectives and research challenges.

2.1. Situation Awareness in Tactical MANETs

The following sections give an insight on situation awareness in tactical MANETs. Firstly, typical application areas for tactical MANETs are presented focusing on disaster recovery. Then the importance and the characteristics of situation awareness in this context are analyzed in more detail.

In the following typical application areas are described for tactical MANETs in the disaster recovery context. Disaster recovery is the implementation of processes and procedures in order to return a system or a society to a state of normality after a disaster occurred. Disasters may be either natural disasters or man-made disasters. There are three essential emergency functions in "normal" civil life:

- *Police* providing community safety and reducing crimes against persons and property.
- *Fire and Rescue Service* providing action forces for firefighting and rescue operations.

- *Medical Emergency* providing personnel and ambulances for out-of-hospital medical care and transport to hospitals.

Additional organization in emergency management and civil defense join forces in a disaster scenario. The relevant organizations in Germany are listed in the following:

- Federal Agency for Technical Relief *THW* (Bundesanstalt Technisches Hilfswerk)
- Federal Office of Civil Protection and Disaster Assistance *BBK* (Bundesamt für Bevölkerungsschutz und Katastrophenhilfe)

For example, the Rapid Deployment Unit Search and Rescue *SEEBA* (Schnelle Einsatz Einheit Bergung Ausland) of the German THW is a tactical unit for urban search and rescue missions. It follows the guidelines of the *International Search and Rescue Advisory Group (INSARAG)* of the UN. The operational units are ready for take-off at the airport within 6 hours after an incident happened. Each tactical emergency squad consists of 15, 32 or 71 members with specific capabilities: search (including search dogs), rescue and medical treatment [Byt05].

In the following a list of examples of natural disasters where tactical mobile ad hoc networks may be used for disaster recovery is given.

- *Earthquakes:* In January 2010 a fierce earthquake hit Haiti [RL10] affecting around three million people. Hundreds of thousands of residences and commercial buildings collapsed or were severely damaged. Communication systems, transport facilities, hospitals, and electrical networks were damaged by the earthquake. Many countries responded with humanitarian aid dispatching rescue teams including medical personnel and engineers. The Tohoku Earthquake in March 2011 was the strongest earthquake in Japan since the beginning of local earthquake records. It triggered two disasters in the region: a 10-meter-high tsunami and several accidents in nuclear power plants in Eastern Japan (e. g., Fukushima) [Bra11].
- *Hurricanes:* In August 2005 Hurricane Katrina was one of the most devastating natural disasters in the history of the United States. It mainly affected the Gulf Coast in southeastern parts of USA and caused an enormous damage. Virtually all communication systems failed during the disaster, including Internet access,

radio communication and cell phones [War05]. In September 2008 hurricane Ike cause a major blackout for 6 million people that lasted between a few hours to over a month [Bla09].

- *Floods:* Levee failures and floods are often caused by hurricanes. Hurricane Katrina caused over 50 failures of levees and flood walls which led to major floodings in New Orleans area. The majority of communication outage after Hurricane Katrina was due to the flooding that prevented refueling of generators [Bla09]. During the presentation of the final report after the flood in Eastern Germany in 2002 the minister of the interior of Saxony-Anhalt criticized frequent network congestion or network outages during the incident. In consequence he proposed project for the improvement in the areas of communication and coordination of forces [Sta03].
- *Tornadoes:* In April 2011 a series of hundreds of tornadoes (2011 Super Outbreak) affected within a few days Alabama, Georgia, Mississippi and Tennessee and several other states. Hundreds of people were killed and it caused property damage of billions of U.S. dollars [UPI11]. The extensive wind damage led to power outages in many areas.
- *Tsunamis:* Tsunamis are caused by undersea earthquakes e. g., outbreak of oceanic volcanoes or submarine landslides. In December 2004 an undersea earthquake occurred in the Indian Ocean close to the west coast of Sumatra, Indonesia. More than 150,000 people in 11 countries died due to the water waves caused by the quake of magnitude 9.0 [Nat05]. The tsunami caused widespread damage of the infrastructure throughout the devastated areas, shortages of food and water, power failures and breakdown of communication infrastructures.
- *Snow:* In November 2005 heavy snow left tens of thousands of households without electricity in northwest Germany and eastern Netherlands. Due to the weight of heavy snow on the lines some power poles buckled or broke; additionally branches fell on power lines. Cable breaks and short circuits led to power outage in wide areas. More than 100,000 people had to spent two nights in subzero temperatures without power. Emergency supplies were set up to heat homes and hospitals [Har05].

There are other natural disasters that cause major damages such as volcanic eruption, pandemic or famine.

Additionally tactical mobile ad hoc networks may be used during recovery operations after man-made disasters.

- *Infrastructure Failures:* In September 2003 a serious power outage for 1.5 to 19 hours affected almost all of Italy [Ber04]. In November 2006 a major blackout in Europe left up to ten million people without electricity. The outage lasted up to 120 minutes in parts of Germany, France, Belgium, Italy, Austria and Spain. The trigger was a planned temporary shutdown of a 380 kV high voltage line for the disembarkation of a cruise ship via the river Ems. People got stuck in elevators, shops were looted and rescue forces were in continuous operation [Spi06].
- *Terrorist Attacks:* The September 11 attacks ("Nine-Eleven") were coordinated suicide bombings where four airplanes were hijacked on domestic flights and crashed on key civilian and military buildings in the United States of America including the World Trade Center in New York City [KHBV*04].

Other man-made disasters include dam failures, explosions, fire, railway crash, nuclear and radiation accidents, spills of oil or other hazardous material.

2.1.1. Situation Awareness

Situation awareness in a tactical environment is based on the gathering, processing and visualization of information concerning the location and characteristics of resources (e. g., people, vehicles, materials) and the environment (e. g., extent of damage, flood extent). The objective is to provide each user with an overall view of the current situation tailored to his capabilities and requirements (cf. [Cou05]). Information can be a force multiplier in tactical environments as knowledge may be of far more value than additional resources (people and equipment).

As described in section 1.1.3 on page 7 situation awareness is (according to Endsley's definition) based on the following three steps:

- *Perception:* Acquisition of (sensor) data related to resources and environment including status, attributes and dynamics, retrieved from the node itself or from other network nodes.
- *Comprehension:* Processing, aggregation and fusion of information to comprehend their relationships and meaning.
- *Projection:* Projection of current state in the near future, e. g., development of the impact of a disaster, development of tactical situation, including expected locations of task forces and victims.

The focus of this dissertation is on state estimation, represented by perception (level 1) and comprehension (level 2), which is the basis for the other steps. State estimation forms the core part which is mostly related and correlated to the specific properties of tactical MANETs. A more detailed analysis of the state of the art of situation awareness approaches and an overview of related work is presented in section 3.1 on page 37.

In the following examples of typical data sets that are acquired, fused and presented to the user for situation awareness in tactical environments are described: starting with knowledge sources, followed by dynamic (sensor) data.

2.1.1.1. Knowledge Sources

Some typical examples of information & knowledge sources in tactical network for emergency response and disaster recovery are described in the following (based on [Cou05, SG11]).

- *Site information:* Description of the location and the surroundings where the disaster happened:
  - Maps/geography: terrain features such as elevations, rivers, shorelines and transport routes
  - Power and water supplies, telecommunications installations
  - Additional information about site location, e. g., digital photos of site
- *Disaster recovery team and resources:* Background information about the disaster recovery team, mission and relevant resources:
  - Team capability and structure, e. g., staffing, chain of command, available specialists

  - Overall mission and team assignments, tactical operational plans, procedures and protocols
  - Available resources (vehicles, devices and materials) for disaster recovery
- *Background information:* Historic and other background information which could further place additional difficulties on the disaster recovery efforts:
  - Information about comparable major disasters, e. g., prior earthquakes in the region and common building construction standards (building materials, common architectures)
  - Safety and security considerations, special hazards, e. g., presence of hazardous materials
- *Logistics and Transportation:* This category covers all aspects of logistics and transportation and partially overlaps with the geography area as transportation routes, e. g., roads, rivers or airways, may also be elements of the geographical knowledge sources.
- *Meteorology:* This area deals with long term data about climate and weather, e. g., seasonal temperature, regional rainfall or snowfall.
- *Equipment:* This category covers equipment items (e. g., devices, vehicles, medical equipment) that is used in tactical operations.
- *Organizations:* This area characterizes various organizations, e. g., research and monitoring organizations, religious organizations and terrorist organizations.

2.1.1.2. Dynamic (Sensor) Data

Typical examples of dynamic (sensor) data sets about the situation in disaster recovery scenarios (based on [SG11, Eur09, ZZNJ02, Cou05]):

- *Disaster recovery team and resources:* Information describing the current status of disaster recovery team, mission and relevant resources:
  - Location of teams and available resources engaged in disaster recovery mission, e. g., specialists expertise, relief supplies
  - Currently ongoing (mission) tasks, planning
- *Incident information:* Description of of the incident and its consequences:

- Description of the incident, type of the incident
- Incident extent, e. g., incident map, flood extent, building damage, gas plumes
- Aggregated victim information: number of casualties, population displacement, disease surveillance, number of detected versus number missing
- Consequences on the environment
- Other relevant information for disaster recovery, e. g., on site gathered information about construction of buildings after earthquakes to analyze cause of building collapse and estimate building stability

- *Victim information:* Information about the individual victims:
  - Victim identification
  - Victim location
- *Other dynamic (sensor) data:* Other dynamic information that may be relevant for disaster recovery efforts:
  - Weather reports and forecasts, meteorological satellite images

2.2. Technical Setup

Based on the discussions above we derive in the following some assumptions about a typical tactical MANET environment and present some mission scenario examples.

2.2.1. Operational Environment

The operational environment of a tactical MANET is determined by the nature of the mission scenarios. Task forces are required to be light, mobile and flexible units with only minimal support assets. A key characteristic is the *absence of a communication infrastructure* as in traditional tactical networks. This is due to the assumption that operations take place in fluid environments where no communication infrastructure is available or was destroyed or should not be used. The terrain and environment may not allow the emplacement of infrastructure components or the exploitation of tactical advantages may not leave time for establishing the requisite infrastructure. There may be, however, limited or intermittent backlinks to a network infrastructure

- *Network Size:* The network contains between 10^1 and 10^2 network nodes. The area of operation (network diameter) is up to several kilometers.
- *Node Mobility:* Nodes are mobile and typically move during most of the operation. This leads to a dynamic network structure and frequent topology changes (available communication links and known routes between nodes). Nodes typically have access to geolocation information (see below) via built-in GPS sensors or via personal area network links (e. g., Bluetooth).

  The movement of individual nodes and the overall group depend on the specific mission scenario and other constraints e. g., due to the environment. Some examples of node movements in typical mission scenarios are given in section 2.2.2 on page 26. A more detailed analysis of research on mobility models and related work is presented in section 3.2.4.1 on page 64.
- *Trust Relationship:* The nodes of a tactical MANET form a closed group of previously known and authenticated participant which is not open to other, unknown nodes. Therefore the initial trust in other network nodes is high. The evolution of trust values should be based on some (cooperative) trust assessment mechanisms that punish suspicious and malicious behavior and lead to increased trust when cooperative behavior and conforming data is gathered.

  A detailed analysis of existing approaches and related work regarding trust and trust assessment is presented in section 3.3 on page 78.

Node Capabilities A tactical MANET is a distributed, decentralized structure where all participating nodes have (at least basic) communication capabilities to jointly form a network. There might be, however, some nodes with higher capabilities ("hubs"), particularly with regard to energy capacity, computational performance and memory and energy capacity, e. g., notebooks or PCs mounted on vehicles.

The emphasis of the discussion of node capabilities in the following is on mobile, hand-held devices because they establish the baseline for available performance to be considered. The challenge is the development of algorithms and modules for context awareness that work on these lower-end devices as well.

There is an increasing desire to use COTS equipment with limited or no adaptations in order to reduce the length of procurement cycles. Therefore the equipment is based largely on high-quality current COTS products with only minimal modifications (e. g., change of maximum power settings or use of high-gain antennae).

- *Radio Interface:* The following provides some typical properties of radio interfaces used in tactical MANETs:
  - Radio interface based on the IEEE 802.11 series of standards (802.11 a/b/g and n on some devices)
  - Bluetooth radio interface for short-range communication
  - Transmission power typically constrained to a maximum of 100 mW, typical power settings between 10 and 20 mW (-10 dBm and -13 dBm)
  - Un-augmented receiver sensitivity ranges from -94 dBm for traffic at 1 Mbps and -71 dBm at 54 Mbps
  - Frequency bands used are between 2.4 and 2.5 GHz and 5.15 and 5.85 GHz
  - Maximum transmission speed is 54 Mbps (with 802.11n theoretically up to 300 Mbps, depending on parameter settings), but typical transmission speeds are below 10 Mbps
- *Computational Capabilities:* The following constraints are relevant in respect to the computational capabilities of mobile devices:
  - Desirable small form factors and therefore limited heat dissipation
  - Peak of computational power is around 4000 to 8000 MIPS and 2.5 to 5 GFLOPS in standard benchmarks
  - Continued computational activities near peak capacity must be avoided due to battery limitations
- *Storage:* The storage capabilities comprise the following components:
  - Main memory
    * 512 MBytes to 2 GBytes for hand-held devices
    * Other nodes (e. g., back-link capable nodes) may hold up to 8 GBytes
  - Secondary storage
    * Hand-held devices typically hold only limited amounts of secondary, non-volatile storage (as hard disk storage may not be robust enough for continued use in tactical environments)

    * Memory cards may hold up to 32 GB of storage, solid-state storage is limited to around 64 GByte (for hand-held devices)
    * Required for storage of additional knowledge sources, e. g., topographic model of the environment, mission information

- *Energy Capacity:* Due to node mobility and device portability the battery lifetime is restricted and battery capacities will probably not significantly increase over the next years as they have only increased slowly in recent years.
  - Hand-held devices are operated with rechargeable or replaceable batteries, individual batteries having a capacity of no more than 5000 mAh
  - Rechargeable batteries offered by third-party vendors may have significantly higher capacities than OEM products are used as baseline
  - Replacement of batteries typically implies shutting down the system, therefore maximum lifetime of an individual battery is a limiting factor, regardless of replacement mechanisms used
- *Sensors:* A node may hold some additional sensors, e. g.,
  - GPS
  - Electronic compass (based on a magnetometer)
  - Accelerometer
  - Temperature, atmospheric pressure and humidity sensor
  - Ambient light sensor

2.2.2. Example Mission Scenario

The following example scenario focuses on providing sufficient levels of detail to allow a discussion and evaluation of specific algorithms and system architectures considering basic requirements and constraints. This is likely to be embedded in a bigger operation and mission scenarios.

Search and Rescue In the following we described a typical search and rescue (SAR) scenario in the disaster recovery context based on the guidelines of the *International*

Search and Rescue Advisory Group (INSARAG) of the UN. A SAR team of the German THW contains the following components [Byt05]:

- Self-sufficiency for 10 days: own food and own water (or water preparation), own camp with toilet facilities, kitchen and workshop
- Self-sufficient radio communication infrastructure (or satellite communication)
- Medical support for team and dogs
- Rescue equipment

Typically there are three different sizes of SAR teams [Byt05, SG11]:

- *Light SAR team:* Team of 15 members for surface search and rescue in the immediate aftermath of the disaster.
- *Medium SAR team (default):* Team of 32 members for technical search and rescue operations in structural collapse incidents, search for entrapped persons.
- *Heavy SAR team:* Team of 71 members for difficult and complex technical search and rescue operations. This category is for disasters where multiple buildings collapsed and is typically found in an urban environment.

A basic step for a tactical SAR team is the reconnaissance and inspection of the affected area and the identification of hazards and measurements to reduce risks. The main task is of course is search and rescue of people using a combination of technical equipment and search dogs. The team needs to take care of rescued victims and applies medical treatment if necessary [Byt05, SG11].

2.2.3. Specific Characteristics of Tactical MANETs

The discussions above show that there are several constraints and characteristics in tactical MANETs that do not typically hold for MANETs in general. The following assumptions that usually apply to typical tactical MANET scenarios are considered for the research challenges addressed in this dissertation:

- *Closed Group:* There is a closed group of participants that are previously known (or at least belong to a specific class of candidates) and typically meet for a briefing before an operation. Therefore the identities of the nodes that belong to the network are well-known and pre-authentication mechanisms (such as shared keys) can be established before the operation.

- *Coalition MANETs:* In tactical MANETs multiple groups and organizations can join their resources to communicate and collaborate for a specific task or mission. In this case each participating party provides its resources to other coalition members and can benefit from the infrastructure and services of the other partners.
- *Specialized Equipment:* The network participants usually use standardized equipment according to well-defined specifications. The equipment is specialized for specific purposes and mission scenarios. Therefore additional external sensors and security mechanisms can be implemented that are not part of the core communication system which is usually considered as the only common denominator for a MANET. This equipment may include also specialized software, applications and databases (e. g., mission profiles or topographic models of the environment).
- *Cooperative Nodes:* All network nodes share a common goal and cooperate. They try to provide each other network node as much and as accurate information about the current situation as possible. They do not behave selfish and therefore also no mechanism to stimulate cooperation is needed, which gives selfish nodes some incentive to cooperate and contribute to network operation. Misbehaving or selfish behavior are only due to an attacker that took over a regular network node.
- *Rigid Policy:* The objective is not to mitigate malicious behavior, but to detect it and to terminate it, e. g., by completely excluding the malicious node.

These characteristics may be exploited to increase robustness of situation awareness mechanisms in tactical MANETs.

2.2.4. Adversary Model

In the following we describe the assumptions about the adversary model. We generally assume that attackers have a least the same capabilities as benign network nodes and the same type of hardware devices at their disposal. In the following we describe some further details about the adversary model:

- *Active Insider Attacker:* Attacks can essentially be divided into external and internal attacks. External attacks are within the scope of this dissertation of secondary importance, since on the one hand they can be repelled more easily and on the other hand they frequently just represent an additional barrier which may be overcome by the aggressor, before he is eventually able to accomplish an internal attack. Attacks which do not require internal access to the network are mostly limited on denial-of-service attacks (DoS attacks), e. g., using a broadband or smart radio jamming attacks, or passive attacks as eavesdropping on network traffic. The latter kind of attack can be addressed using cryptographic protection mechanisms.
  Internal attacks are accomplished by attackers, which are legitimate participants of the network and exploit their trustworthy position. These may either be saboteurs or more likely adversary forces which have compromised a regular node of the tactical MANET. The attackers have all necessary authentication and encryption material at their hands and thus are part of the network. A malicious node could, for example, send false routing information saying it has the most recent (highest sequence number) or shortest route to a certain node, although this is not true, in order to reroute all traffic through this malicious node (cf. [EB06, BETJ06]). Furthermore there is no need to gain physical access to a network node to join the network, this way attacks can be executed from any location within radio range.
- *Multiple and Cooperative Attackers:* There may be several attackers in action simultaneously that work cooperatively. Attackers may join their efforts and cooperative attack a selected victim from different positions within the network in various ways at the same time.
- *Byzantine Behavior:* Our adversary model allows Byzantine behavior of attacking nodes, i. e., they may deviate from normal behavior in an arbitrary way. In particular, instead of not contributing to the network communication and application protocols these nodes try to maliciously disturb accurate assessment and attempt to alter the system to an inconsistent state by inserting false values. Fur-

thermore Byzantine nodes may collaborate to amplify the impact of their attacks and to avoid detection. Additionally the attackers do not constrain themselves to a malicious and therefore conspicuous behavior avoiding detection. Instead they try to attack and disrupt the system in the most effective way. However, we do not expect that normal network nodes behave selfish. Misbehaving or selfish behavior are only due to an attacker that took over a regular network node.

- *Attack Objectives:* An attacker may pursue different objectives such as saving resources, disrupt network performance or gain tactical advantage. The latter is of most interest within this dissertation, e. g., determining the location of other node(s) or gathering other sensitive information. A specific object of an attacker may be to partition the network into at least two sub-networks: nodes from two different sub-networks cannot communicate with each other although a route between these nodes actually may exist. Less important within this context are denial-of-service attackers disrupting network services or selfish nodes that want to save their own resources by forcing other network nodes to take more of the workload, e. g., forwarding of network traffic.

2.3. Objectives

In this section we specify the overall objectives of this dissertation to increase to provide robust situation awareness in tactical MANTEs particularly focusing on design objectives, and research challenges.

2.3.1. Situation Awareness in Tactical MANETs

The objective of situation awareness in tactical MANETs is to *enable and sustain coordination.* The team members should be provided with a common operating picture showing the current status, actions and potential impacts. The goal is to understand what is currently happening in order to plan what the next steps. In disaster recovery this means to manage the incident and to maintain operations in order to support victims and foster fast recovery.

Situation awareness should enable tactical forces to *share information* and collaborate. This includes the integration of information from distributed and disparate sources. Therefore robust information distribution is required supporting various kinds of data including video, images and other complex data sets.

The situation awareness system needs to analyze and integrate multiple pieces of information in order to understand their meaning and *comprehend the current situation*. Therefore the relationship and relevance to personal and mission objectives has to be determined and potential impacts assed.

The results should be *presented to the user* as a common operating picture showing forces, resources, their current status and potential impacts of the current situation. This is typically a map-based user interface showing dynamic sensor data, images, text messages and other documents. However, cognitive overload of the operators has to be avoided. Therefore this application has to be tailored to operational needs of specific missions, minimizing information loads while still delivering all necessary information.

The focus of this dissertation is on entities composing the actual tactical MANET: task force members and their equipment, e. g., communication devices and vehicles. These components represent the particular challenges, peculiarities and opportunities for situation awareness in tactical MANETs. Within the scope of tactical MANETs the core challenge is to assure security and safety of the operational team and the successful completion of overall mission objectives.

2.3.2. Design Objectives

In the following design objectives for robust situation awareness mechanisms are specified.

- *Accurate Situation Assessment:* We require a concept that thoroughly captures the current state, analyzes the current situation and detects misbehavior of other network nodes. We require a state estimation concept that is highly accurate and provides a good estimate of the current network and node states.
- *Misbehavior Detection:* Malicious behavior has to be unambiguously detected and high detection rate have to be achieved while the number of false-positives

is low. The system should be even robust against a large number of cooperating, malicious nodes.

- *Graceful Degradation:* We require a scheme that can cope with faulty behavior of network nodes or (partial) failures of system modules. The negative effects of a degradation of the amount and quality of information that is available should be minimized.
- *Cooperation:* We require a decentralized architecture that allows the nodes to cooperatively assess the current situation and detect malicious behavior.
- *Efficiency:* We require a protocol that minimizes communicational overhead, i. e., the number of additional packets to be send and the overall amount of data to be exchanged.

2.3.3. Research Challenges

Security measures to increase the robustness of situation awareness should not focus on solving specific problems, e. g., detection of a specific attack type, as new attacks will arise that was not thought of before. Active protection can only be a first part of an overall security concept; a more general approach for robust situation awareness is needed.

Therefore the objective of this dissertation is to *provide robust situation awareness in tactical MANETs* by addressing the presented research challenges (cf. section 1.2 on page 10). We pursue the following objectives:

- *More effective utilization of available data sources* to counter false or malicious data, e. g., considering overlapping data sources and cross-relationships.
- Detection of *potential threats to the SA system*, e. g., injection of false data, and cooperative trust assessment in order to be resilient to malicious nodes.
- *Robust and flexible system state modeling and assessment* since measurements and data are error-prone.

We exploit and utilize communication protocol extrinsic data sources, i. e., additional sensor data and knowledge sources, cooperation mechanisms and state estimation. The focus of this work is less on exploiting routing protocols and communication systems as these areas have been addressed by many research groups and thoroughly analyzed.

The main objective is increasing robustness in state estimation for situation awareness in tactical MANETs. This objective is broken down into three research challenges represented by the following three questions (see also section 1.2 on page 10):

1. How can location-related *data sources* be utilized in a *comprehensive and effective way* to detect malicious behavior?
2. How can the *trustworthiness* of other nodes be *efficiently and cooperatively* assessed?
3. How can *system states* be *modeled in a generally applicable way* and *comprehensively assessed* to counter inaccurate and error-prone data?

Technological solutions need to meet the specific challenges by potentially highly skilled and powerful attacker within the given application scenario.

Utilizing location-related data sources in a comprehensive and effective way to detect malicious behavior The objective is to provide a *comprehensive assessment of the current situation.* Therefore, models and mechanisms are required that enable the exploitation of all available data sources. False or malicious data should be detected in order to increase the robustness of state estimation for situation awareness in tactical MANETs. Not only device characteristics and communication protocols should be considered but also application layer information and additional knowledge sources need to be incorporated.

For these purposes, a *general model and analysis for the broad set of available data sources* in tactical MANETs is required. This needs to include data provided by dynamic direct or indirect sensor data as well as static mission-specific or general knowledge sources (e. g., mission profiles, mission profile, tactical mobility patterns, topography model). Local data sources as well as data provided by other nodes needs to be incorporated. Redundancies within all these data sources need to be discovered caused by partial overlaps and cross-relationships. Effective mechanisms need to be developed for modeling and detection of contradicting data sets in observations and knowledge sources.

Node mobility and related aspects pose particular challenges in MANET environments. Therefore, mechanisms for *comprehensive and effective analysis of location-*

related data sources for detecting malicious behavior are required. Node locations directly influence and correlate with other situation characteristics: Task force tracking, topography of the environment as well as technical system aspects such as network communication. Cross-relationships of node positions, routing data and other data sources need to be investigated in order to detect potential inconsistencies. Particularly the challenges for the incorporation of radio signal characteristics have to be addressed as radio signal propagation highly depends on the topography of the environment.

Efficient cooperative assessment of the trustworthiness of other nodes Establishing a common situational picture in tactical MANETs is based on the cooperation with other network nodes. Therefore, the *trustworthiness of information provided by other nodes* needs to be analyzed. An efficient and effective assessment of the trustworthiness of other nodes is the basis for robust cooperation. The special characteristics of tactical environments necessitate the development of specifically tailored cooperation mechanisms for distributed trust assessment. In tactical environments all participants share a common goal and mission and there is a sharp borderline between benign nodes (that are a legitimate part of the network) and malicious nodes (attackers). The developed mechanisms need to be particularly robust against attacks or node failure.

A *robust and efficient communication model* has to be developed for distributed cooperative trust assessment in tactical MANETs. The model needs to consider resource constraints of mobile devices and the special communication characteristics of wireless ad hoc networks. These characteristics provide challenges, such as lossy transmission channel, as well as potentials, such as broadcast communication.

Trust assessment of individual nodes is based on a combination of local misbehavior detection with information provided by other nodes. The *modeling of trustworthiness information* needs to enable the effective incorporation of indirect trust assessments. Corresponding and conforming judgments should be considered and incorporated as well as contradicting assessments. The trustworthiness model should reflect the certainty and the accuracy of particular trust estimations. Mechanisms are required that allow the incorporation of the local trust assessment with trust estimations provided by neighbor nodes.

Modeling of system states in a generally applicable way and comprehensive state assessment to counter inaccurate and error-prone data For robust situation awareness a *probabilistic state modeling and estimation* is required that provides robustness in respect to inaccurate and error-prone data. Measurements may be inaccurate, faulty sensors may lead to false observations and/or attackers may intentionally inject wrong data into the situation awareness system. The developed mechanisms should be generally applicable and allow for flexible incorporation data sources.

The system state and state propagation model should be open and *applicable to a wide range of application scenarios*. State modeling should include discrete as well as continuous states. The state propagation model should be also applicable for nonlinear systems and Non-Gaussian process noise. This way the state assessment mechanisms are not restricted to specific application scenarios but generally applicable for situation awareness in tactical MANETs.

State estimation mechanism are required that enable flexible exploitation of available data sources. They should allow for *comprehensive state assessment* based on the incorporation of multiple observations and additional information sources. This should include observations and estimations provided by other network nodes based on a distributed state estimation architecture.

The following three chapters describe how the objectives can be reached by addressing the research challenges described above by developing suitable mechanisms for cross data analysis, cooperative trust assessment and probabilistic state modeling.

3. Related Work

In this chapter we analyze and discuss research concepts and technologies that can be utilized to increase the robustness of situation awareness in tactical MANETs. In the following section the state of the art regarding situation awareness in tactical MANETs is analyzed. We present some definitions for situation awareness (SA) and discuss SA concepts for cooperative and distributed environments, in particular the handling of faulty and/or maliciously behavior.

Afterwards related work is presented in respect to the three research challenges addressed within this dissertation: cross data analysis, cooperative trust assessment and probabilistic state modeling. We discuss the modeling of cross-relationships and consistency checks and the utilization of location-related data sources for the detection of malicious behavior. Then we present concepts for trust modeling and mechanisms for trust assessment and cooperation. Finally, we discuss probabilistic state modeling focusing on particle filters and the particular challenges of cooperative state assessment in distributed environments.

3.1. Situation Awareness

There are several related cognitive concepts that should not be confused with situation awareness. Since there are no clear definitions for all of these concepts there is still a lack of clarity and some ambiguity about terms.

Situation awareness (SA) can be seen as a state of knowledge that is achieved as a *short-term* objective. The processes used to achieve SA are called *situation assessment* [End95]. Therefore situation assessment can be defined as the *"process of achieving, acquiring, or maintaining SA"* [End95]. As described already in the introduction *state estimation* is a core part of the situation assessment process.

3.1.1. Situation Awareness Definition

There have been numerous attempts to define the term *situation awareness (SA)* [Roe00]. In the following some of the more prominent definitions are presented.

- Sarter and Woods define SA as the *"accessibility of a comprehensive and coherent situation representation which is continuously being updated in accordance with the results of recurrent situation assessments."* [SW91]
- Vidulich *et al.* define SA as the *"continuous extraction of environmental information, integration of this information with previous knowledge to form a coherent mental picture, and the end use of that picture in directing further perception and anticipating future events."* [VDVM94]
- Smith *et al.* define SA *"as adaptive, externally directed consciousness"* [SH95]

As presented in section 1.1.3 on page 7 the most common and generally accepted definition of *situation awareness* was phrased by Endsley, which is also used within this dissertation:

"Situation awareness is
the perception of elements in the environment within a volume of time and space,
the comprehension of their meaning, and
the projection of their status in the near future." [End95, EC08]

This means that all information that is available should be gathered, analyzed and aggregated to create an awareness of what is happening around a particular network participant and to project the impact of the current situation on the near future.

Endsley's SA model is composed of three steps or levels (see Figure 3.1): perception, comprehension, and projection. These three levels are described in more detail in the following [End95].

- *Perception of Elements in Current Situation (Level 1)* This is the basic level of SA that refers to the perception of all relevant elements (e. g., objects, persons, events, environmental factors) in the environment. It includes the monitoring of

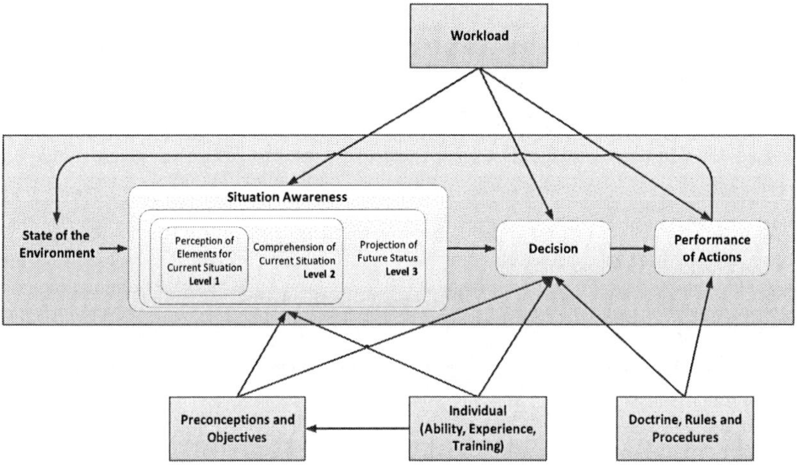

Figure 3.1.: Situation Awareness Model based on Endsley [End95, BP02]

their current states (e. g., locations, status, attributes), and their dynamics. On this level simple recognition and cue detection processes are implemented.

- *Comprehension of Current Situation (Level 2)* The objective of the next level of SA is to understand the meaning of elements in the environment. This involves processes for the integration and synthesis of information related to disjoint (level 1) elements. Patterns are recognized, interpreted, and evaluated in order to understand how the current situation will impact on own mission goals and objectives. The outcome should be a comprehensive picture for the user that shows all relevant parts of the environment. For example, a tactical commander needs accurate information about all task forces within a given area, e. g., type, location, number, capabilities, and dynamics.
- *Projection of Future Status (Level 3)* The objective of the third level of SA is to predict changes and future states of elements in the environment for a sufficient period of time. The basis for the projection is a thorough knowledge of current state and dynamics of elements and the comprehension of the situation (level 1 and 2 SA). This assessment of the current situation and a model of the overall system are used to extrapolate information forward in time in order to predict what

is likely to happen next. The results have to be presented to the user in a suitable way to facilitate SA. Based on the situation assessment responsive actions are triggered, partially automatically, e. g., for testing and verifying an attack, but mainly based on user judgment.

Endsley defines SA explicitly as a construct separate from decision making and performance of actions. For example, best-trained and experienced decision makers may take wrong decisions if they are provided with inaccurate or incomplete SA. In contrary, users who have perfect SA may still make wrong decision if they lack relevant training or experience. There are several additional factors that influence the overall process. This includes the overall workload, preconceptions and objectives as well as individual characteristics (ability, experience, training) and general doctrines, rules and procedures.

Time and Space For SA both *temporal and spatial aspects* are of particular importance. Although SA consists of a user's knowledge of the state of the environment at a given point in time, it is highly *temporal* in nature as in dynamic systems the situation is always changing. Therefore the assessment of the current situation must constantly be updated; otherwise it will be outdated and thus inaccurate. Information for SA can not necessarily be acquired instantaneously, but has to be built up over time in order to be able to capture system dynamics [End95, End00].

SA is also highly *spatial* in many contexts. Spatial as well as functional relationships among elements are useful to determine which aspects of the environment are relevant for SA. Only information that is relevant to specific tasks and mission objectives should be considered. In a tactical environment the relevance of other elements will typically depend on their location and relative speed [End95].

Team Situation Awareness In tactical scenarios typically several individuals may work together as a team to carry out tasks or missions. Typically multiple team members share common SA requirements, i. e., information requirements overlap. Team members should operate on common data sets as the assessments and resulting actions of one team member can have a crucial impact on others. Team members of a poorly functioning team may derive inconsistent situation assessments based on shared SA requirements and therefore behave in an uncoordinated manner that may be even counterproductive. On the contrary team members of smoothly functioning team share a joint understanding of all common SA elements, this is this purpose of shared SA [EJ97].

Shared situation assessment may lead to several possible states:

- Two team members share the same *correct* SA.
- Both team members have a *wrong* SA.
- Two team members have *different* SA, where one has the correct and one an incorrect view.

The goal of shared SA is that both team members share the correct view on the situation.

Different pictures of the same situation may be easily revealed. If two team members detect contradicting views they can start a process to gather additional information or work together on resolving the differences. The most dangerous situation occurs of course when both team members share a common but incorrect SA. In this case there is no contradiction, the problem will not be detected and the team members will continue the operation based on wrong SA [EJ97].

Team SA comprises a collective awareness of the situation, team members must possess SA related to their individual roles and goals and SA related to other team members. A simple team SA model was proposed by Salmon *et al.* [SSW*08, SSWJ09] that contains three components: individual SA, team processes and team SA (see Figure 3.2). Salmon *et al.* [SSW*08, SSWJ09] argue that this simple concept may be only sufficient for simple, small-scale collaborative scenarios but not applicable for complex, large-scale collaborative scenarios.

Distributed SA Distributed SA (DSA) [SSW*08,SSWJ09] considers team SA not as a shared understanding of the situation. DSA is an entity that is rather separate from individual team members, it is distributed within the system and a characteristic of the overall SA system [AG98, SSW*08]. DSA comprises team members as well as the artifacts that they are using (see Figure 3.3). Artman and Garbis define team SA as: *"the active construction of a model of a situation partly shared and partly distributed between two or more agents, from which one can anticipate important future states in the near future."* [AG98]

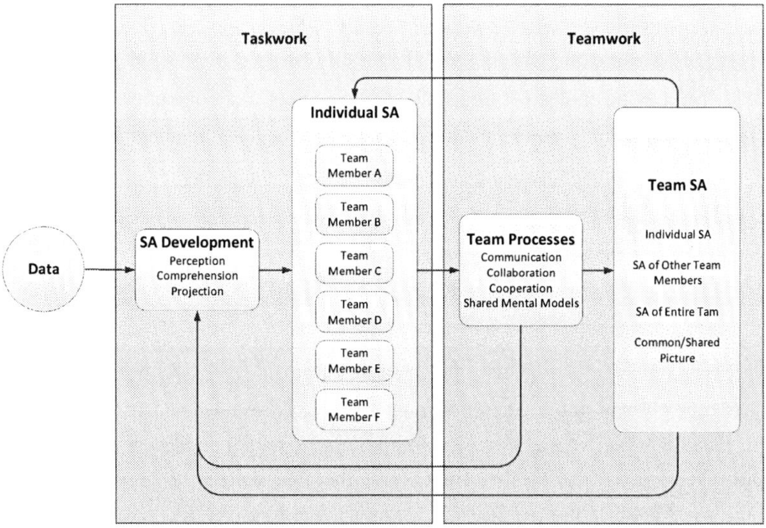

Figure 3.2.: Team Situation Awareness Model [SSW*08, SSWJ09]

The literature review of Stanton *et al.* [SSW*08] on both individual and team SA shows that existing related work is diverse and inconsistent. Individual and team models of SA are, however, distinct from DSA approaches. An important aspect is the nature of the concept, whether it should considered as cognitive or as systems construct. Stanton *et al.* conclude that in collaborative systems systems-oriented approaches are most suitable, e. g., the approaches presented by Artman and Garbis [AG98] and Stanton *et al.* [SSH*06].

3.1.2. Errors and Cyber Threats in Situation Awareness

The problem of false, missing or malicious data in SA have been addressed and discussed by several research groups. Initial work was mainly restricted to system failures, e. g., false or missing sensor data and human errors, e. g., due to information overload. Later work also considers intentionally falsified or inhibited data by attackers.

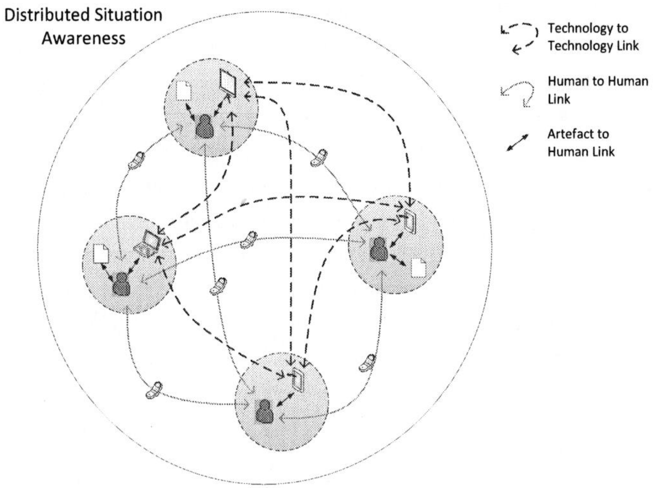

Figure 3.3.: Distributed Situation Awareness Example [SSW*08]

Errors in Situation Awareness There are several reasons and multiple types of failures in SA when only incomplete or inaccurate and erroneous knowledge is available for some elements of the environment. The factors that can seriously affect SA effectiveness are described in the following for the different SA levels [End95]:

- *Level 1 SA errors* The main errors on this level are failure to perceive important information for SA regarding a specific task and misperception of observed information. Missing information leads to incomplete SA and may be due to the lack of detectability of the (physical) characteristics of a specific element or event or to a failure in the system design that hides this information from the user. A person may become aware of this problem with a coinciding error. For example, a person may not realize that it is icy only until he or she slips. In other cases of incomplete SA the relevant signals could be observed by the operator, but they are not properly perceived. Humans have several general weaknesses in sampling such as the misperception of statistical properties, the visual dominance over audio information and the limitations of human memory and attentional narrowing. Normal sampling strategy may have been interrupted due to some external events and not reactivated in a timely manner. Misperception may be due to ambient lighting, e. g., seeing a blue light as green or seeing a 3 as an 8.

- *Level 2 SA errors* Level 2 SA errors are based on an inability of the user to properly integrate or comprehend the meaning of perceived data. There may be several reasons for this: the necessary mental models may not be available, a wrong model was selected or pieces of received data may be mismatched with the model. When a wrong model is select based on situational cues this model is used to interpret all perceived data, which will lead to wrong integration and comprehension of the data. Even when the correct model was selected, however, errors may occur. Received data may be mismatched with the existing model or not matched at all. When new situations are encountered for which known default values are not appropriate, a new model needs to be developed or an existing model is adapted in order to include this new class of situations.
- *Level 3 SA Errors* There might also be errors or failures on level 3 SA. For the projection of future dynamics a highly developed mental model is required. Even if a situation is well assessed understood, it may be impossible to accurately project future dynamics without a suitable mental model.

A problem is the *detection of SA errors* as users are often not aware of what they do not know and the actual level of precision of their picture of the current situation. Erroneous SA may be detected when a person receives some new data that does not fit with expectations. This conflict may be resolved by revising the existing model, developing a new model, or by changing task plans in order to bring perceived data and system model in line. A wrong choice, however, can easily render SA efforts ineffective for some time. When data is perceived that does not fit into the existing model, a common problem is to continue to decide whether the existing model should be further revised or if a new model should be chosen or developed.

Cyber Threats on Situation Awareness Endsley and Jones [EJ01] propose concepts for modeling the impact on situation awareness of cyber attacks as well as interruptions and information disruptions that are not due to some hostile actions. They argue that the effects on SA and decision making may be very similar in both cases. Cyber attacks may become more common place in tactical scenarios as terrorist activities increase.

Users need SA support for effectively detecting and dealing with information attacks and normal disruptions.

They define four categories of disruptions based on affected processes and matters [EJ01]:

1. Information preprocessing
2. Prioritization and attention
3. Information confidence level
4. Information interpretation

Disruptions that affect information preprocessing may be naturally occurring within a specific operation environment (benign) or due to malicious attacks. Increasing the overall amount or the rate of provided information may slow down information processing and lead to an incorrect SA picture. Users may omit key information if multiple sources provide disagreeing information. Dissonant information does not necessarily indicate malicious activities; it may also be, for example, due to differences in sensor technologies. *Disruptions affecting prioritization and attention* may be, for example, due to interruptions of the users by new competing tasks, alarms, or other events in general directing her attention towards specific elements. This way the user may focus in on the interrupting task and (temporarily) forget about other competing goals and tasks. *Disruptions that affect confidence in information* may be very difficult to detect. Attackers may partially corrupt some information in order to make certain information sources look unreliable (compromised or faulty). Low confidence in specific information (sources) may have significant negative effects on SA. In this case the SA and decision making processes may be ineffective even if correct information is available. *Disruptions affecting interpretation* may be based on insertion or withholding of information. Attackers may insert cues that make provided information look consistent with known "normal" situations and lead users to interpret it as a wrong class of situation. On the other side critical information may be withheld that would indicate a different class of situations. When an attacker managed to make a user activate a wrong situation model it may be very difficult to detect the actual attack.

Choobineh *et al.* [CAG12] propose a system to measure the impact of cyber threats on mission activities and resources. They use the process modeling notation BPMN

(Business Process Model and Notation) in order to model and measure the impact of security breaches on processes and used resources.

Jakobson [Jak12] goes one step further by introducing the notion of cyber attack tolerant missions. His objective is to increase the tolerance to cyber attacks against information assets supporting mission operations. The proposed system is based on federated adaptable and situation aware multi-agent systems. The idea is to increase security of current and future mission operations by dynamically adapting to new cyber security situations.

Errors and cyber threats pose are generally a challenge for situation awareness. This challenge is even bigger in dynamic and distributed tactical environments.

3.1.3. Situation Awareness in Tactical MANETs

In the following we present approaches for SA in tactical environments in the area of disaster recovery. The goal of SA is to provide users with an overview of incoming information, an operational picture of the current situation and potential future events (including their likelihood) in order to support decision making during operation.

The objective of the OASIS project [Cou05] was the development of a framework that can serve as a basis of a European Disaster and Emergency Management system. On the tactical level they propose to use a MANET communication infrastructure for firefighters, police and medical staff [Cou08]. They state that SA is a key factor for the effectiveness of disaster and emergency operations. Since SA relies on exchanging and compiling collected information they defined a standard for the description of tactical situations, the so called Tactical Situation Object (TSO).

One of the requirements for tactical SA within the OASIS project was the detection of misbehaving nodes. They do, however, not solve this problem and leave it as an open issue as they claim that current state-of-the art in misbehavior detection is not yet suitable for these purposes [Cou08].

The SSMC/DDKM (Seamless Services and Mobile Connectivity in Distributed Disaster Knowledge Management) project [SLLJ09] aims to develop a distributed disaster knowledge management system. The objective is that shared situational awareness between participants in emergency scenarios can improve reaction ability and accelerate decision processes. The project examines how mobile technologies can be utilized within this context to effectively support disaster recovery operations.

The goal of the SoKNOS (Service-Oriented ArchiteCtures Supporting Networks of Public Security) [PPBS10, PDTS*09] project was to develop concepts for supporting organizations and companies in emergency scenarios. Service-oriented solutions enable ad hoc information sharing and the flexible integration of heterogeneous information sources. All actors are provided with a common view of the operational situation on a number of different devices including mobile devices that are carried in the field.

DistressNet [GZC*10] is an architecture based on wireless ad hoc and sensor networks for supporting disaster response. Several mechanisms are combined, e. g., distributed collaborative sensing, topology-aware routing, and accurate resource localization, to provide first responders with detailed situational awareness

General Concepts The state estimation and SA approaches for tactical environments presented above are mainly based on the following components.

- *Mobile communication network:* radio network for mobile connectivity, partially based on MANETs
- *Data integration:* data categorization, indexing and linking, data aggregation, data fusion
- *Data reduction:* filtering of relevant and irrelevant or duplicated information in order to prevent information overload
- *Knowledge and information management:* distributed mechanisms for management and dissemination of situation information
- *Security mechanisms:* protection of critical information, monitoring of the situation for threats, no counter-measures against false or malicious data
- *User interface:* visualization & interaction, e. g., geospatial map-centric system aiding situational awareness by displaying resource locations and messages

The presented concepts neglect or only marginally address potential risks and threats to state estimation and SA systems, e. g., sensor failures or injection of false data by malicious nodes. The data modeling and analysis concepts do not utilize all available information to detect and increase robustness against wrong or malicious data. Overlapping data sets and cross-relationships should be used to detect errors or inconsistencies. They use explicit data and system models for state estimation which are not flexible and robust enough to deal with measurement based and error-prone data are. Data is taken "as is", aggregated and fused in order to derive a specific value for each parameter.

3.1.4. Detection of Malicious Behavior in MANETs

Intrusion Detection Systems (IDS) are directly related to state estimation and SA. They also analyze and collect sensor data and process it with the objective to assess the current situation, particularly focusing on the detection actions that may compromise the system.

An IDS is generally a system, which detects attacks to a computer network and/or on a computer system, similarly to an alarm system for monitoring areas or buildings. An IDS passively monitors its environment on malicious behavior and/or states and creates an alert whenever suspicious situations occur. IDS and related concepts may a building block for increasing robustness of SA systems. Location based IDS will be discussed in detail within the section about Cross Data Analysis (see section 3.2.4.5 on page 75).

Classical Intrusion Detection approaches in wired networks can be distinguished on the one hand by the objects that they are monitoring in network and host based systems, and on the other hand according to the method of the recognition process in signature and anomaly-based intrusion detection systems. A *Host based IDS* runs on a computing system (host) and collects, monitors and analyzes data of the internals [GZH*12]. A *Network based IDS* collects network traffic, either on the network interfaces of a computer or within the network, e. g., on hubs or routers. Their focus is the detection of malicious network activities, such as denial of service attacks or port scans.

Signature based IDS (also known as rule based IDS) identify intrusions by watching for patterns of traffic or application data supposed to be malicious. It analyses information gathered and compares it to large databases of attack signatures. Essentially, the IDS looks for a specific attack that has already been documented. As a virus detection system, the IDS is only as good as the database of attack signatures that it uses to compare packets against. These types of systems are presumed to be able to detect only known attacks. *Anomaly-based IDS* monitor the activities of a system and tries to classify the behavior in either normal or anomalous. The classification is based on heuristics of normal system operation. Therefore anomaly based IDS may detect any type of misuse.

The structural and behavioral differences between wired networks and MANETs render existing IDS designs not suitable for MANETs. Therefore new concepts and methods have been developed and proposed in the literature. *MANET IDS* approaches may be categorized into four main areas as described in the following.

- *Standalone IDS* [BB03a] run on each network node independently. There is no cooperation and no exchange of (sensor) data or alert information. Therefore detection decisions are based only on locally collected information. Due to its limitations this architecture is not very effective as local information might not be sufficient to detect attacks. This approach isnot very popular and may only be applied if not all nodes are capable of locally running an IDS or for other reasons an IDS should not be installed on each node.
- *Distributed and collaborative IDS* [ZL00, ZLH03] execute an IDS agent on each network node that participates cooperatively in the intrusion detection process. Nodes may exchange (sensor) data and alert information. When local detection mechanisms are inconclusive a cooperative global IDS assessment may help to clarify the situation. This is a very common and popular architecture for MANET IDS:
- *Hierarchical or cluster based IDS* [SBC*05, SWP03, HL03] is a specific case of a distributed and collaborative IDS. It is suitable for multi-layered network infrastructures where the MANET is divided into clusters, e. g., using a corresponding routing protocol. Cluster heads may have more capabilities than cluster members and act as control points similar to routers classical networks. On each cluster member an IDS agent is run that is locally responsible for monitoring and intrusion detection. The IDS agents of cluster heads are additionally globally responsible for its cluster, e. g., initiating a global response when an intrusion is detected.
- *Mobile agent based IDS* [AC02, KG03] uses mobile agents to perform specific task on other network nodes on behalf of the agent owners. This allows flexibly

assigning and distributing intrusion detection tasks within the network. This may reduce computational complexity as some functions need not to be performed on every node. A mobile agent can freely move within the MANET in order to perform one specific task. The system is scalable in large and varied system environments and tolerant against network partitioning. Mobile agents must, however, be protected from execution environment on remote hosts.

These concepts are not independent and may also be combined. A mobile agent based or a hierarchical or cluster based IDS is also a distributed and cooperative IDS. A mobile agent based IDS may, for example, be combined with a hierarchical IDS.

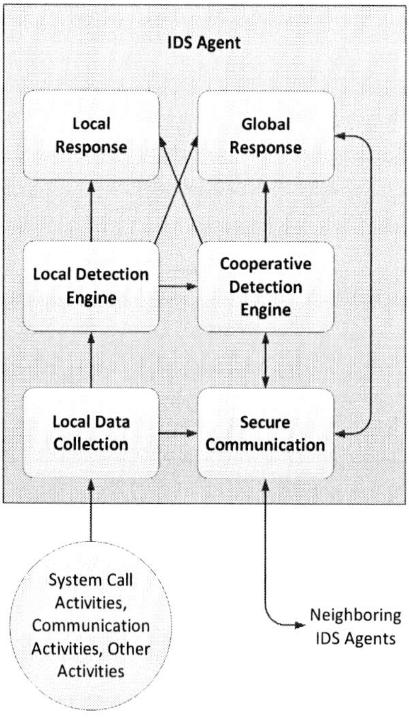

Figure 3.4.: Architecture of a Cooperative MANET IDS [ZLH03, ZL04].

A very generic model for distributed and collaborative MANET IDS was proposed by Zhang and Lee [ZL00, ZLH03] which can be used for most cooperative intrusion detection approaches and therefore was adopted by many researchers. The proposed architecture for the IDS agent is composed of six modules as shown in Figure 3.4.

- *Local Data Collection* This module collects real-time data from various sensors. These sensors can gather data from user and system applications; and from network packets, including those observable within the radio range of the monitoring node.
- *Secure Communication* The secure communication module provides reliable and secure communication infrastructure between the individual IDS agents.
- *Local Detection Engine* This engine processes the data collected looking for intrusions. At this step, not only misuse detection techniques should be used but also anomaly detection, as it is very probable that new attack types will be developed.
- *Cooperative Detection Engine* When a node detects an inconclusive intrusion, it starts a cooperative intrusion detection process. This consists of broadcasting the information about the potential intrusion to the rest of the nodes. If other nodes find enough evidence, a response is triggered. An important difference between local detection and cooperative detection is that in the first case information is analyzed from the local node only, in the second case an IDS agent relies also on data from other agents.
- *Local Response* The local response modules triggers local response actions. This could be, for example, triggering an alert to a local user.
- *Global Response* The global response coordinates reactive actions among neighbor nodes depending on the type of intrusion, e. g., network protocol. For example, a response could be to re-initialize communication channels (e. g., force re-keying) or identifying compromised nodes and re-organizing the network in order to exclude them.

This generic scheme for cooperative IDS was subsequently extended, e. g., by [MNP04].

Several approaches address the problem of selfishness, i. e. nodes that use the network but do not cooperate. A typical approach is to categorize nodes based upon their behavior as proposed Marti *et al.* [MGLB00]. A watchdog is used to identify misbehaving nodes and a pathrater helps routing protocols to avoid these nodes.

Buchegger and Le Boudec [BB02a, BB02b] describe a mechanism to detect and isolate misbehaving nodes based on "grudger nodes" called CONFIDANT (Cooperation Of Nodes: Fairness In Dynamic Ad-hoc NeTworks). Trust relationships and routing decisions are based on experienced, observed, or reported routing and forwarding behavior of other nodes. For this purpose each nodes generates and disseminates alarm messages when malicious behavior was directly detected or reported by other nodes. Incoming alarm messages are filtered based on the trust level of the reporting node and for trustworthy alarms a respective reaction is triggered. The concept comprises four components: a monitor for neighborhood watch, a reputation system that manages a table of nodes and their rating, a path manager, and a trust manager that deals with incoming and outgoing alarm messages.

Michiardi and Molva [MM02] propose a collaborative reputation mechanism to enforce node cooperation (CORE) for MANETs. The behavior of other network entities is observed and the reputation of nodes is based on their willingness to collaborate in order to counter selfishness. Only positive values are assigned to avoid denial of service attacks. Direct observations are combined with information provided by other network nodes. They argue that reputation is compositional and that the overall opinion on a node is composed of subjective reputation, indirect reputation and functional reputation.

Bansal and Baker [BB03a] propose mechanisms for observation-based cooperation enforcement in ad hoc networks (OCEAN) avoiding a complicated trust-management machinery by using only direct first-hand observations of other nodes' behavior. Neighbor nodes are watched and ranked according to their behavior. The route finding and packet routing process is based on this ranking and traffic from misbehaving nodes is rejected. A timeout-based second chance mechanism allows nodes that were previously considered misbehaving to become useful again if they change their behavior.

The basic idea of [YML02] is to introduce tokens to every network node which enables its participation in network activity. Local neighbours are responsible for mutual verification of tokens. At the beginning tokens expire quickly such that neighbours have to often re-verify tokens. The behaviour of nodes is overheard and checked by adjacent nodes. As long as nodes are working according to the network protocol the expiration time for tokens is increased. If nodes in the 1-hop neighbourhood detect misbehaviour, the tokens of the accused node will be revoked for all nodes in the network, such that intruders are directly excluded from the MANET.

[XPS11] provides an overview of MANET IDS architectures which may be adapted and applied within tactical environments.

However, these approaches focus on IDS only and not on SA. In the following section we present related work on cross data analysis in tactical MANETs. We discuss how location-related data sources can be utilized for the detection of malicious behavior in order to provide comprehensive and effective SA.

3.2. Cross Data Analysis

In this section we provide an overview of related work to cross data analysis. We discuss concept and mechanisms that can be used to comprehensively and effectively utilize location-related data sources for robust situation awareness in tactical MANETs. We start with a general introduction of data fusion and give an overview on cross-layer approaches for wireless mobile networks. Then we describe plausibility and consistency checks of data sets and give some background information about cross data analysis of location related data.

3.2.1. Data Fusion

In the following we provide an overview and some background information on data fusion. It generally involves the combination of data to estimate or predict the state of some entities (in the past, at current time or in the future), for example, identity, activity, location, movement or other attributes. When the state estimation involves people it is important to also consider their informational and perceptual states – of individuals as well as of groups. Additionally interaction with other physical objects and prediction of future developments have to be addressed [HL01].

The following definition of data fusion (also known as multi-sensor data fusion) is given in [HL01]: *"Data fusion is the process of combining data or information to estimate or predict entity states"*. Data fusion provides an important functional framework for SA as it requires the fusion of data from many heterogeneous distributed sensors [Bas00] .

Over the years, many data fusion models have been proposed but the most generally referenced and used as the model proposed by the Joint Directors Laboratory Data Fusion Working Group (JDL). As shown in Figure 3.5 [SHB04] the JDL model comprises five levels: Sub-Object Data Assessment (Level 0), Object Assessment (Level 1), Situation Assessment (Level 2), Impact Assessment (Level 3), and Process Refinement (Level 4) [HL01].

While Endsley's model for situation awareness model (cf. section 3.1 on page 37) is based on the human perspective, the JDL model provides a functional model for the

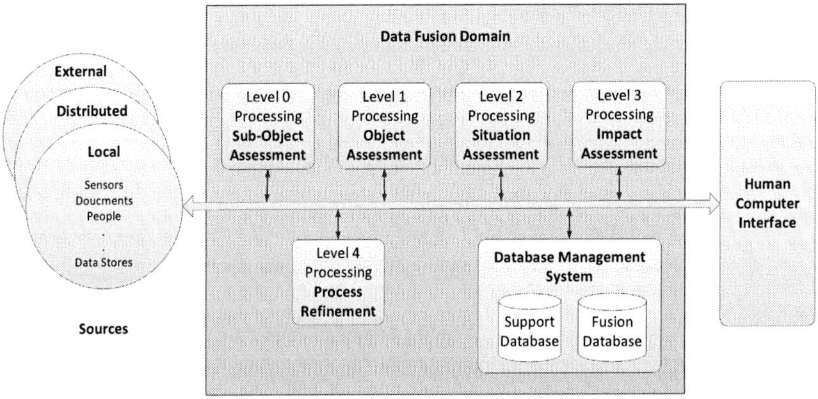

Figure 3.5.: JDL Model of Data Fusion [HL01, HM04]

data fusion process. In some respects both can be seen as alternative models based on different viewpoints [SHB04].

Salerno *et al.* [SHB04] analyze various existing techniques that are required for situation awareness. They combine Endsley's situation awareness model with the JDL data fusion model and defined an overall system architecture for situation awareness. They further investigate the role of level 2 and level 3 of the JDL data fusion model and the logical boundary between these levels [Sal07, Sal08]

3.2.2. Plausibility and Consistency Checks

There are multiple interrelated and partially overlapping data sources available in tactical MANETs that can be utilized for situation awareness. The idea of cross data analysis is to detect inconsistencies across these data sources based on plausibility and consistency checks.

These concepts have so far been mainly explored in the context of data warehouses and data warehouses. In the following we present related work about plausibility checks and data inconsistency that may be used to detect incorrect data due to node and sensor failures or attacks.

3.2.2.1. Plausibility Checks

A *plausibility check* describes a validation mechanism whether a value or a result is plausible and therefore acceptable. This can be done for example by roughly estimating (at least the magnitude of) the measured or calculated values in order to check whether the measured value is reasonable, i. e., matches the estimate, and therefore is acceptable. The meaning of *plausibility* within this context is something related to consistency, accuracy or credibility. Plausible data seems to be valid, likely, acceptable, credible, comprehensible or even convincing.

The basis for the implementation of plausibility checks is existing knowledge about the system and the environment. Plausibility checks can be useful to detect obviously incorrect values. For example, if the result of the calculation of the average height of people in a country, e. g., Germany, is 2.5 meters, plausibility checks may easily show that this value cannot be right since this value is not plausible. The advantages of this approach are the low implementation cost, a disadvantage is, however, that less obvious mistakes may not be recognized.

Plausibility checks are carried out, for example, during credit approval processes by a bank on the data that is provided by the credit applicant or is available by other means. During this process as much as possible relevant data is collected, checked for completeness the plausibility of the available data is assessed [TN04].

An important application area for plausibility checks is database management and data warehousing. Before data is imported and integrated into a database system the data needs to be checked in order to ensure a high data quality. Business process management systems may apply plausibility checks to ensure the validity of the steps that are performed [SAP08, SAP13].

The following gives an overview on various types of methods for plausibility checks and data validation (cf. [SAP08, Mic11]).

- *Data type and encoding:* It should be examined whether the data type of the input data matches with the expected data type, e. g., a character string or a numeric field.
  - Allowed characters

  - Data format and structure
  - Validity of codes
- *Numerical checks:* There are several kinds of numerical plausibility checks and data validation methods.
  - Control and batch totals
  - Cardinality
  - Check digits (or check sums)
  - Range and limits
- *Data consistency:* Simple consistency checks may verify that data in one fields corresponds to the data in other fields of a data record, e. g., if the title of a person is "Mr." the gender should be male.
- *Cross system consistency:* More elaborate checks may examine data from different data sources or even in different systems and check whether it is consistent, e. g., name and gender of a specific user with the same identification number in several systems.
- *Complex validation:* Even more complex checks might be possible when complex interdependency are modeled and considered as well. The evaluation may need to incorporate policy limits and temporal aspects in order to accomplish this kind of checks, e. g.to model and verify validity of individual transactions if there is a limit on the accumulated costs per month.

The latter three types of methods for plausibility checks are all *consistency checks* between more than one data sources. These data consistency mechanisms are more elaborated and will be further discussed in the following section.

3.2.2.2. Data Inconsistencies

Data consistency describes the integrity of related data based on the validity of some known relationships. Another approach is to look at data inconsistency as the presences of some contradictions or conflicts within the available data sets. Stuller defines *inconsistency* in data warehousing and *inconsistency data* in the following way [Stu99]:

Definition 3.1. *"A database has an* inconsistency *if the data it contains yield under the given interpretation at least one contradiction."*

Definition 3.2. *"The concrete data of a given database which yield a contradiction will be called* inconsistent data."

Inconsistencies in data warehousing may be classified according to Stuller based on their sources and the stage of the data warehouse processes [Stu99]:

- *Conceptual inconsistencies:*This kind of inconsistencies is addressed during schema integration, a new conceptual model or a redesign of the logical schema of the data warehouse might be required.
- *Semantical inconsistencies:* This problem should be addressed during the integration of the databases, in particular by a deliberate choice of the attribute(s) to be considered.
- *Data inconsistencies:* Data inconsistencies should be addressed at the data entry stage by a thorough verification and validation of the data to be added.

The detection of any type of these inconsistencies may give important feedback for conceptual modeling and logical schema design, improve data entry, data verification and data validation and help choosing most suitable data mining techniques [Stu99].

Greenfield [Gre11] names inconsistencies as one out of four categories of "errors" that may occur during building process of a data warehouse (incompleteness, incorrectness, incomprehensibility, and inconsistency). Inconsistencies include the use and meaning of codes, information aggregating, graining of most atomic information, inconsistent timing and out of synch data.

As *incomplete and inconsistent UML models* cause problems in the software development process inconsistencies were also investigated within this context. In [LCM*03] Lange *et al.* propose the use of inconsistency and incompleteness metrics to predict the quality of resulting software systems. They differentiate between two characteristics that may be present in a collection of UML diagrams that refer to one specific software system: contradicting information (inconsistency) and missing information (incompleteness). They define *consistency* as the *"compliance between different views or diagrams of a design"*.

Inconsistencies in Knowledge Management In the field of knowledge management the term *consistency* of knowledge is often defined as the absence of contradictions. Nguen [Ngu08] criticizes that in this case the notion inconsistency is simply replaced by the notion contradiction. In contrary he argues that the following three components are required to describe an inconsistency of knowledge:

- *Subject:* The part of the real world that the inconsistency refers to.
- *Set of elements related to this subject:* Knowledge elements, for example, a relational tuple in the knowledge base.
- *Definition of contradiction:* The criterion of contradiction of a set of formulae.

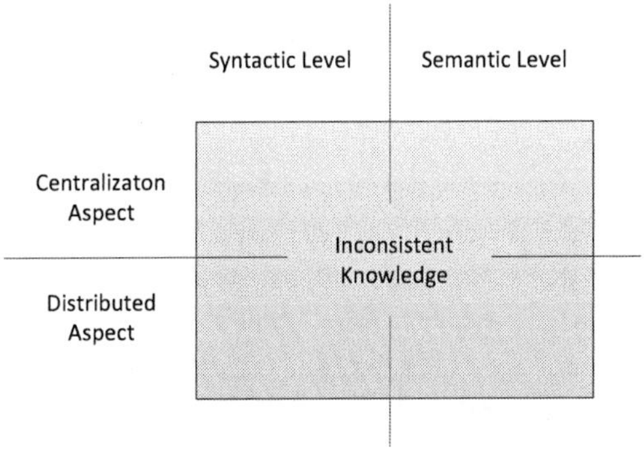

Figure 3.6.: Levels and Aspects of Knowledge Inconsistency [Ngu08].

As shown in Figure 3.6 knowledge inconsistencies may be categorized according to [Ngu08] in two levels (syntactic and semantic) and divided into two aspects (centralized and distributed). The *centralization aspect* characterizes inconsistencies of knowledge within one knowledge base. The *distribution aspect* reflects inconsistencies between different knowledge bases. The latter kind of inconsistencies arises in distributed environments when several systems generate different conflicting versions of knowledge referring to the same subject.

A practical example may be a multi-agent system for SA in a distributed environment (cf. [Ngu08]). A set of nodes run monitoring agents that monitor their neighborhood consisting of several neighbor nodes They observe, process the captured data and draw conclusions to assess the current situation. The monitoring areas of agents mutually overlap. When a specific event of interest occurs each agent analyzes the available data and deduces the kind, source, and propagation of this event. The deduction results, however, may be inconsistent. Different agents may diagnose different characteristics for the same event and the same node, i. e., there may be a conflict.

The example above shows that conflicts arise when several cooperating nodes generate different versions of data regarding the same subject. Conflicts occur when (at least) two entities have different opinions regarding the same subject. Nguyen [Ngu08] proposes a formal model for the representation of conflicts that consists of the following components:

- *Conflict body:* participants of the conflict, e. g., a set of agents.
- *Conflict subject:* matters of the conflict, a set of issues.
- *Conflict content:* opinions (knowledge states) of the participants regarding the conflict subject.

The concept of inconsistency is also addressed within the field of logic. In contrary to classical and other logics paraconsistent logics are tolerant to inconsistencies. Numerous approaches (e. g., [KL92,Kni02,Hun03]) have been proposed to formalize inconsistent but non-trivial theories. Within paraconsistent logics measures have been proposed that quantify the inconsistency value based on the sets of formulae which are not consistent. The first formal model for conflict was presented by Pawlak [Paw98] where positive knowledge was represented by "+" and negative knowledge by "?". There are, however, some differences in the semantics of Pawlak's "neutrality" and the semantics of Nguyen's "uncertainty". Neutrality in a voting process does not mean uncertainty but uncertainty means incompetence to express its opinions on some matter.

3.2.3. Cross-Layer Approaches for Wireless Mobile Networks

In the following we present a review of cross-layer approaches for wireless mobile networks as they are related to the cross data analysis approach developed within this dissertation. However, the focus of these approaches is on optimizing network operation [FGA08] and not on situation awareness.

Cross-layer approaches have become a popular research topic in network communication research. They break with the typically used and well-established communication concepts based on separated layers and allow information sharing between different layers. The general goals are to increase the knowledge about the current state of the network and to optimize communication based on new and more efficient protocols. Combining various co-related problems (such as routing, scheduling and channel assignment) can increase the overall performance. Cross-layer design can enable new distributed, simpler and provably efficient algorithms [PD11].

It is often argued that layered architectures serve well for wired networks but that they are not as suitable for wireless networks [SM05]. The main idea of layered communication architectures is to divide the networking task into several layers, typically based on the seven-layer open systems interconnect (OSI) model [ISO94]. Each of these layers provides specific services and communication with the corresponding layer at the other communication endpoint based on specific protocols. The main restriction is that communication is only allowed between directly adjacent layers based on predefined interfaces. The main advantage of this layered approach is that communication protocols can be developed in isolation and exchanged independently at specific layers.

The reason for violating the layered communication architecture by developing a cross-layer design is to exploit dependencies between protocol layers in order to obtain performance gains. Three main reasons for cross-layer design in wireless mobile networks are due to the specific characteristics of the wireless medium [SM05]:

- *A specific problem that cannot be handled well by layered architecture frameworks.* For example, packet transmission errors on a wireless link may be mistaken in a layered TCP/IP design as an indicator for network congestion. The typically triggered response mechanisms are not suitable to efficiently and effectively address the problem.
- *Wireless transmission enables opportunistic communication.* The potential of these possibilities cannot be sufficiently exploited in layered design. An example is the opportunistic usage of a channel due to time-varying link quality.
- *New communication modalities based on wireless transmission.* These new modalities are not accommodated by layered designs. For example, the physical layer

in wireless communication may be capable of receiving multiple packets at the same time.

Srivastava and Motani [SM05] the define cross-layer design as *"protocol design by the violation of a reference layered communication architecture ... with respect to the particular layered architecture"*. They also present a taxonomy for the basic types of violating a layered architecture by cross-layer design:

- Creation of new interfaces
- Merging of adjacent layers
- Design coupling without new interfaces
- Vertical calibration across layers

These violation types may also be combined in more complex cross-layer designs.

[FGA08] provide a survey and taxonomy of cross-layer approaches for wireless mobile networks. Their focus is on optimized operation of mobile devices in modern heterogeneous wireless environments. They describe four areas of problems that can be addressed using cross-layer design: security, mobility, quality of service, and adaptation of the wireless link. For example, a security module may coordinate encryption across layers. This way unnecessary duplication of encryption on several layers can be avoided, reducing processing efforts and related power consumption in order to improve network performance. They describe the major open technical challenges in the cross-layer design research area and point out that there are ongoing efforts for the integration of cross-layer design solutions into wireless communication standards.

Besides performance optimization of network communication based on cross-layer design there are also some approaches for security solutions addressing multiple network layers. Kidston *et al.* [KLTM10] present a cross-layer security service framework for tactical networks integrating all communication layers. Security services at upper layer (application layer) may use sensing data from lower layers (physical or data link layer). They present a lightweight integrated authentication (LIA) scheme as an example how cross-layer design enhances security and reduces communication overhead between nodes. In subsequent work [KLAML11] Kidston *et al.* present a cross-layer protocol for cluster based data dissemination. The proposed scheme improves the efficiency and effectiveness of a detection mechanism for disconnected nodes.

König *et al.* [Kön06] present a geographically secure MANET routing mechanisms based on a cross-layer approach. Their main idea is to prevent the routing of confidential information through areas within the reach of unauthorized (and potentially malicious) persons. Routing itineraries are restricted based on specific security requirements. This way provides the proposed solution long-term protection for confidentiality by avoiding information disclosure.

Kawadia and Kumar [KK05] provide a critical view on cross-layer design and show some problems and unwanted effects. They state that there is always a tension between high performance and good architecture. Therefore, developers may be tempted to violate architectural principles in order to optimize performance. However, performance improvements in cross-layer design may go along with a loss of good architectural design reducing proliferation and longevity. In their view architecture should be considered as performance optimization over a longer period of time.

They present some examples how unintended cross-layer interactions may have negative impact on overall system performance. For example, cross-layer design may create loops or may lead to spaghetti design hindering further innovations. Careless cross-layer design may result in code that is difficult to maintain and extend. They also provide some simulation studies showing how cross-layer design may lead to unintended interactions and negative impact on system performance.

The presented cross-layer approaches try to optimize the system in respect to a specific objective by breaking the existing architectural design and protocols. In contrary the goal of cross data analysis is to exploit all available data in order to get a more comprehensive view on the current situation without interfering and breaking architectures and protocols.

3.2.4. Cross Data Analysis of Location Related Data

Location related data sources are used as the application scenario for cross data analysis within this dissertation. Therefore related work in the field of cross data analysis of location related data is presented in this section. We start with a review of mobility models as they are important to model node placement and movement. Afterwards we discuss radio propagation models proposed in the literature as basic technology for utilizing radio signal characteristics. Distance measurement using radio signal strength or signal propagation times is a promising approach for cross data analysis of location related data. Finally, mechanisms for consistency checks of location and routing data for attack detection are presented.

3.2.4.1. Mobility Models

The movements of the users have a significant influence on the behavior of the network, which is why accurate modeling of user motion is of particular significance (see also [BB06]). Hence mobility models have great influence during simulations on network topology and dynamics. Firstly, node positions are dynamically determined based on these models which network connections are accordingly available. Secondly, they directly influence how quickly existing network connections are broken and therefore new routes must be established. Therefore the success and the detectability of a MANET attack depend to a big extent on node mobility. The selection of the right mobility model has a major impact on simulation results [THB*02, CBD02]. The following review of MANET mobility models is not exhaustive; rather it should show what approaches exist and what developments are going on.

Random waypoint mobility models In random-based mobility models [CBD02] node movements are determined by a small number of parameters (such as the variance of a Gaussian distribution) and are limited by a few constraints (e. g., area of movement). One of the most widely used probabilistic models is the *Random Waypoint model* [Wil04, CBS03, JM96] which is basically the simplest model for user movement. Nodes randomly determine their initial positions, which are determined by a uniformly distributed probability function. Each node randomly selects a new destination in each simulation cycle. The result is basically some kind of Brownian motion which means that users are gradually distributed over the entire motion area. Since the nodes in this model, always follow the shortest path to their destination, the node density in the middle of the simulation area is higher than in outlying areas.

Path based and topography model based mobility models Random waypoint mobility models do not consider the environment of the user, they are rather assuming that users can move freely on an open area. Due to this they can only imprecisely reflect real world aspects. The goal of graph-based mobility models is to simulate realistic mobility constraints of MANET users. An example of this type of mobility models was developed by Stepanov *et al.* [THB*02]. The core idea is that MANETs users in the real world usually follow terrain elements such as roads.

In graph-based mobility models, for example, the streets of a city are modeled as edges and the intersections represent the nodes of a graph. Users always move on the edges between nodes. The result is a mobility model, in which users behave much more realistically than in the previously presented models.

The *Manhattan Grid Model [Spe98]* is based on road topology. Roads are located in a grid structure and nodes move in horizontal or vertical directions on these roads. The model follows a probabilistic approach: each node probabilistically chooses at each intersection to keep moving in the same direction or to turn left or right. An implementation of the Manhattan Grid Model is available in BonnMotion [AEGPS10]

Jardosh *et al.* [JBRAS03, JBRAS05] introduced a mobility model that is based on a topography model. In this *Obstacle Mobility Model* buildings are modeled as polygons and the radio transmission between two nodes is interrupted, or at least greatly mitigated, if a polygon obstructs the direct line-of-sight between two network nodes. Nodes can either move on predefined paths or they try to reach their destination on the shortest path through the obstacle field.

Activity-based mobility models The activity-based model mobility developed by Stepanov *et al.* [SHR05] is an extension of the graph-based model. The influence of properties of the environment on the behavior of nodes shall be considered even more than it is the case in graph-based models. For this reason, two new concepts were introduced: activity and user trip. An *activity* represents a typical user activity in the real world that an agent can perform at a certain place such as shopping. A *user trip* is a sequence of edges in the spatial model graph. In general, each user trip is associated with an activity trip, i. e., the agent will randomly select an activity and creates a user trip that takes him to a node at which this activity is offered. This model is more realistic than the graph-based model because it provides a more realistic distribution of agents.

Group mobility models The previously described models handle the mobility of individual nodes independently and therefore do not consider any group mobility behavior. In group mobility models, however, each node usually moves relative to the center of a defined group where the movement of the center of the group is based on one of the models described above. Group mobility models are generally more difficult to implement and less explored.

The first group mobility model for MANETs, the so-called emph Reference Point Group Mobility (RPGM) Model, was introduced by Hong *et al.* [HGPC99] in 1999 to represent the relationship among a mobile nodes. When RPGM each group has a defined center and the nodes move randomly but uniquely distributed in certain distances from this center. Wang and Li [WB02] extended the RPGM in their *Reference Velocity Group Mobility Model (RVGMM)*, in which the movements of the nodes is depending on the speeds of the other nodes in the group.

In [BL04] Blakely and Lowekamp presented the *Structured Group Mobility Model (SGMM)* where nodes stay in formation, i. e., the relative position of nodes in the group is fixed in respect to the group center. This mobility model may be applied to situations where groups move with a a priori known structure such as firefighters operating in a building.

The mobility model proposed by Williams and Huang [WH06] is probably the first that combines group mobility with the presence of obstacles. They build on RPGM and introduce repulsive forces that push agents away from obstacles and other agents in order to avoid collisions with other nodes or obstacles.

Scenario based mobility models for tactical networks Mobility models specifically tailored for tactical MANETs are presented by Aschenbruck *et al.* [AFMT04, AGPM08, AMC11]. The goal is to deliver a realistic modeling for disaster areas in civil protection.

In such scenarios action forces are strictly structured and organized: group leaders specify who has to move where to do what based on tactical reasons. The operations area is separated into different areas: incident site, casualties treatment area, transport zone, and hospital zone. The proposed mobility model was implemented and evaluated for two real-life disasters. Guo and Huang [GH08] present a mobility model for first responders that move within a large disaster area. The disaster area is split into small squares and each group of first responder continues working in a specific square until it is cleared. The basic movement pattern is as follows: when a square is cleared by a group of first responders they split and move on to the adjacent uncleared squares – the more work is to be done in a square, the more first responders enter a specific square.

Aschenbruck *et al.* [AGPM08] for disaster areas. The proposed mobility model is goal driven and supports various types of mobile and stationary units. The model supports group mobility where groups of nodes may move in formations throughout a realistic environment model which may contain obstacles. The model can be applied to various tactical application scenarios.

Schwamborn *et al.* [SAM10] present a trace-based approach where movement traces are statistically analyzed and used for parameterization of a new realistic mobility model. The analysis is based on operational data from the real world scenario. The mobility model for first responder scenarios also considers geographic restrictions based on a topography model of the environment.

Aschenbruck *et al.* [AGPM08] present a survey on mobility models for tactical mobile networks. They evaluate a multitude of mobility models according to different dependencies and requirements for tactical scenarios:

- *Temporal dependencies:* node movement is influenced by previous movements.
- *Spatial dependencies:* node movement is affected by other nodes in the neighborhood, e. g., group mobility.
- *Geographic restrictions:* nodes movement is restricted to some predefined areas.
- *Requirements for tactical scenarios:* heterogeneous velocity, tactical areas, optimal paths, obstacles, units leave the scenario, group movement.

Additionally, a comprehensive survey of movement traces is presented by Aschenbruck *et al.* [AMC11] including an analysis and modeling in order to derive synthetic mobility models. They conclude that although several researchers have analyzed traces their accuracy is often still limited. They claim that there is a huge demand for future work in this area, in particular for capturing traces and generalized analysis. Interdisciplinary research may help to improve results in this area and to develop new models.

3.2.4.2. Radio Propagation Models

Distance estimation based on radio signal characteristics is relevant for cross data analysis of location related data sources within this dissertation. Therefore, we present an overview of the most prominent radio propagation models described in the literature in the following.

- **Free Space Model:** The *Free Space model* is the most simple model and forms the basis for all more comprehensive models [Bal02]. It describes the spherical propagation of radio signals without any disturbances and is based on the assumption of a free first Fresnel Zone. The *Fresnel zones* are concentric ellipsoids of revolution which define volumes in the radiation pattern of (usually) circular aperture. Therefore it assumes the existence of an unaffected line of sight between sending and receiving node.

  Based on these assumptions, the model calculates the power P_r received by a node r at distance d from the sending node, which is qualitatively given by $P_r(d) \sim 1/d^2$. More precisely the receiving power at a receiving node in this model is [GW98, Rap02]

  $$P_r(d) = \frac{P_t G_t G_r \lambda^2}{(4\pi)^2 d^2 L} \quad , \tag{3.1}$$

  where t and r are indices for sending and transmitting node. G_t and G_R describe antenna gains in case of beam antennae and are generally set to 1, since there is usually no use for beam antennae in ad-hoc networks. L ($L \geq 1$) is an additional factor, to take consideration of a power loss caused by the system hardware, which is also set to 1 by default. P_t is the transmitting power, λ the wavelength and $(4d\pi)^2$ describes the surface of a sphere with radius d.

- **Two-Ray Ground Model:** Another basic propagation model is the *Two-Ray Ground model*. It is an extension of the Free Space model that additionally takes the reflection of the radio wave on the ground into consideration. Several experiments have shown that radio propagation in the GHz range is dominated by interferences waves following the direct line of sight and waves that are reflected by the ground (cf. [BOP06] [FV11]).

Receiving power at a receiving node r at a distance d is given by

$$P_r(d) = \frac{P_t G_t G_r h_t^2 h_r^2}{d^4 L} \quad , \tag{3.2}$$

where h_t and h_r are additional parameters describing antenna heights. The wavelength and surface of the sphere are not taken into consideration anymore; instead, the model simply assumes a power decrease proportional to $1/d^4$. This quartic decrease is caused by the assumption of horizontally polarized radio waves, which yields a phase shift of the reflected wave and thus finally a power reduction.

Equation (3.2) is suitable only for distances greater than a threshold distance d_c. For distances smaller than d_c, the Free Space model from eq. (3.1) should be used. The threshold distance d_c is evaluated [Rap02] by equating eq. (3.2) and eq. (3.1):

$$d_c = \frac{4\pi h_t h_r}{\lambda} \quad . \tag{3.3}$$

Both signals (direct line of sight and reflection at ground) have a phase difference and interfere with each other. This results in periodic attenuation of the signal strength in respect to the distance between sender and receiver, particularly in the near field. These effects are considered in the following more sophisticated equation for the Two-Ray Ground model ([GW98]):

$$P_{Rh} = \left(\frac{\lambda_0}{4 \cdot \pi \cdot d}\right)^2 \cdot P_T \cdot G_T \cdot G_R \cdot 4 \cdot \sin^2\left(\frac{k_0 \cdot z_T \cdot z_R}{d}\right) \quad , \tag{3.4}$$

Where P_{Rh} is the signal strength of the horizontally polarized radio signal in Watt, d the distance between sender and receiver. λ_0 is the wave length, P_T is the transmitting power, G_T and G_R are the antenna gains of sender and receiver, respectively, k_0 is the wavenumber and z_T and z_R are the height of transmitting and receiving nodes above the ground. An additional equation for vertically polarized signals is not required since vertically and horizontally polarized signals generally behave in the same way [GW98].

- **Shadowing model** Free space and two-ray ground model calculate receiving power using a deterministic function, which means that radio connectivity would abruptly end at a certain distance. In reality, receiving power values at a specific distance are scattered by time-dependent interference and scattering effects.

  The *Shadowing model* tries to incorporate interference parasitic effects which results in varying radio range and receiving power values. The equation for calculating the receiving power of this model consists of two parts and is qualitatively given by [FV11]

$$P(d) \sim 1/d^{\beta} \cdot X \quad , \tag{3.5}$$

  where $0 < \beta \in \mathbb{R}$ provides a configurable parameter for adjusting the parasitics and X is a random variable modelling scattering.

  The first part is a deterministic function which includes parasitic effects:

$$P_r^1(d) = P_r(d_0)\left(\frac{d_0}{d}\right)^{\beta} \tag{3.6}$$

  $P_r(d_0)$ is the receiving power at a specific distance d_0 close to the transmitter antenna, which must be experimentally determined by measurement. The shadow factor β depend on the environmental setup, e. g., a value of 2 for an free space outdoor environment and 2.7 to 5 in a shadowed urban area [FV11]).

  The second part describes the variation of the receiving power using a $N(0, \sigma^2)$-distributed random variable X, i. e., X is normally distributed with mean 0 and standard deviation σ. Typical values for σ are between 4 and 12 for an outdoor environment [FV11].

- **COST Walfish Ikegami** The *COST Walfish Ikegami model* [Bal02] considers obstacles (e. g., buildings) in the vertical plane and effects such as multiple diffraction over rooftops between the transmitter and the receiver node. The transmitter node is assumed to be 4m to 50m above the ground and the distance between nodes needs to be at least 20m. This model is therefore mostly constrained to environments in which the transmitter is located on a rooftop or similarly elevated terrain feature.

- **TIREM** The *Terrain Integrated Rough Earth Model (TIREM)*[1] is an advanced proprietary propagation model. It was developed for the Joint Spectrum Center (JSC) of the Department of Defense in the US and comprises a set of physics-based, DoD standard algorithms. It incorporates reflection and diffraction effects on topographical objects, tropospheric scattering as well as absorption by the atmosphere. TIREM may be used to predict communication coverage ranges for mobile radios and sensor acquisition ranges.
- **Ray-Optical Propagation Model** A *ray-optical propagation model* that takes reflection and diffraction effects off buildings into account was introduced by Hoppe *et al.* [HWL99]. The model requires the pre-processing of the environment, which by far exceeds the computational capabilities of mobile devices. Pre-processing the data on a powerful server and then storing it on the mobile device is infeasible due to the size of pre-processed data. The use of this model is therefore restricted to the use of powerful computers or to devices where storage or sending of large volumes of data is possible. The accuracy of this model was verified in [RWH02], showing the potential of the approach to model radio propagation by a ray-optical model. The model was further extended by Rautianen *et al.* for the use in indoor environments in [RHW09].

The former radio propagation models are more simple and can efficiently be implement. However, they provide lower accuracy. The latter, more sophisticated models are typically optimized for high accuracy and not intended for use in a resource-constrained environment in which computation must be completed within a near-real-time interval, minimizing energy consumption in the process, and with potentially arbitrarily constrained memory resources.

3.2.4.3. Distance Measurement Using Signal Strength

An application of the presented radio propagation model is distance estimation based on received radio signal strength measurements. Signal strength may be directly read from common Wi-Fi or Bluetooth components. For known transmission power, antenna characteristics, and possibly a topography model of the environment an estimate

[1] `http://www.alionscience.com/en/Technologies/Wireless-Spectrum-Management/TIREM`

of the distance between the transmitter and receiver can be derived from the received signal strength. Most of the works in this area, however, refer to the localization of mobile nodes using an existing infrastructure of base stations (access points) and not on networks without infrastructure, such as MANETs.

Commercial off-the-shelf (COTS) WLAN cards (IEEE 802.11) provide the application developer with the Received Signal Strength Indicator (RSSI), which can be interpreted as a radio signal strength in *dBm*. Due to the fact that wireless cards were not designed for accurate signal strength measurements, the implementation of RSSI varies from card to card and calibration is required. Andersen *et al.* [ARY95] proposed a complex propagation model of signals based on a topography model that can be used to obtain realistic estimates for received signal strengths at a certain distances. The attenuation of signal strength is highly dependent on the environmental conditions and there are also other effects that should be considered, e. g., directional characteristics of antennas.

In this context suitable and realistic radio propagation models are required in order to estimate radio signal characteristics with sufficient precision. In the past, several attempts have been carried out to determine radio signal characteristics in different environments, from rural to dense urban areas to investigate and to model, eg. [RAOR91] and [XMLSR93]. However most experiments on signal strength measurements and radio propagation model verification were carried out in cellular networks where mobile devices communicate with base stations. In [BOP06] similar experiments were carried out for radio communication based on IEEE 802.11b to investigate the two-ray ground propagation model.

Guo [Guo12] present a presents a comparative performance analysis of three location algorithms based on signal strength measurements using an array of sensor. Their location approach addresses non-cooperative wireless intruders in a campus scenario. Their results indicate that no location algorithm always performs better. The localization results depend very much on the constellation, whether the wireless intruder is inside of a circular or grid formation of sensors or outside of the sensor constellation.

Islam *et al.* [IRHS12] propose an approximation technique for node locations in tactical MANETs. Their approach is based on link-state routing protocol data (e. g., OLSR) and a set of anchor nodes or known landmarks. They assume that transmission ranges are approximately known in advance. The accuracy of their method increases with the number of number of anchors and landmarks that are available. In the future they plan to integrate signal strength and propagation time measurements to further enhance their algorithm.

A potentially more promising option for the future is distance estimation based on propagation delays, which are presented in the following section.

3.2.4.4. Distance Measurement Using Signal Propagation Times

Different types of signals could be used for distance estimation based on signal propagation time. However, using the already present radio signals of the communication interface may be the first choice as this causes only little overhead and cost. Challenging is the fact that radio signals propagate at speed of light, i. e. approximately 300 000 m/s, which means that they travel around 300 meters in one microsecond. Therefore, very precise clocks are required for the measurement of radio signal propagation times. An introduction to and overview of the possibilities for distance determination by means of signal propagation times is given in [KBR97].

In general, three methods for distance estimation based on signal propagation times should be considered:

- *Determining the signal propagation time (Time of Arrival, TOA)* Measuring the signal propagation is theoretically the most simplistic approach for estimating the distance between two mobile nodes. This requires, however, very precise and highly synchronized clocks in both network nodes. If such clocks are present one node simply sends a radio pulse and then informs the receiver at which point in time this pulse was sent. When the radio pulse reaches the recipient, it stores the arrival time t_r and waits for the message from the sender about the time t_s when it started the sending of the pulse. The time difference is the signal propagation time $\Delta t = t_r - t_s$. Since the radio signal propagates with a known finite speed v, the distance s can be calculated according to the following equation: $s = v \cdot \Delta t$. The use of radio signals for TOA distance estimation is problematic due to the high precision and synchronization requiremts for the clocks. Therefore, this method is in practice currently not yet suitable for use with small mobile devices.

- *Determining the difference of signal propagation times (Time Difference of Arrival, TDOA)* While for distance determination based on TOA both, sending and receiving nodes, need to have accurate and synchronized clocks, this is not the case for distance determination based on TDOA. For this method it is sufficient if the sending nodes are provided with accurate and synchronized clocks.

  Although the TDOA method is the basis for position determination using GPS it cannot be exploited for alternative distance measurement methods in tactical MANETs since it requires an infrastructure. The primary application area is the absolute position of mobile devices with the help of base stations, whose positions are known and their clocks synchronized. Base stations send messages with current timestamp and from the differences of the signal delays the mobile node can determine its position. For propagation time measurement only the first arrival of a radio signal is considered and reflections [VK04,AO05] are neglected.

- *Determining the overall signal propagation time for a round trip (Round Trip Time, RTT)* Another way of determining the distance between two mobile nodes using signal propagation times is the measurement of the time that a signal needs for a round trip from one to another node and back. This time difference is known as the so-called round-trip time Δt. Ideally, the distance of two nodes s would be calculated as follows (where v is again the propagation speed of the radio signal):

$$\Delta s = \frac{v \cdot \Delta t}{2}$$

  This equation would apply if the radio signal would be directly reflected by the recipient. Since the receiver needs to process the signal and can return a response signal only afterwards, there will be an additional delay and the RTT increases. If the duration of this processing time is known and identical each time, the equation may be adjusted by a constant time which needs to be subtracted from the measured RTT. If the duration of the processing procedure is not predictable the results are incorrect and therefore the distance estimations have a high variance.

It is desirable, of course, to use (standard) components that are already present for distance estimation. The main advantage of distance determination based on the overall RTT is that, in contrast to TDOA or TOA, no synchronized clocks are required. A very accurate measurement of time differences is however still necessary, e. g., as deviations in the range of microseconds represent an estimation error of several hundred meters for the propagation distance. Sufficient accuracy may be already achieved in a stationary experimental setup based on statistical methods.

Guenther *et al.* [GH04,GH05] explain different approaches to perform measurements of signal propagation times with clocks of standard Wi-Fi chipsets. Despite the delays due to signal processing and the fact that these clocks do not have a resolution in the nanosecond range they achieved an accuracy of about 8 meters.

Therefore, distance measurements using standard components based on IEEE 802.11b are generally feasible. In [RGS*06] was shown that the maximum precision of distance estimation based on delay measurement using IEEE 802.11b is under ideal conditions in the range of centimeter. In real-world environments with multi-path propagation an accuracy of 4 to 12 meters was achieved. Thus, an estimate of the distances between each network node on the signal propagation is possible and therefore alternative distance estimation methods can be used to complement geo-coordinates determined by GPS.

Within the AmbiSens project [SWV*08, VSH*08] radio propagation times are used for range estimation of robots. The proposed approach is called two-way TOA, however, it is actually based on RTT measurements. The proposed two-way TOA system is based on the property of IEEE 802.11 that each receiver immediately replies to a successfully decoded data packet with an ACK packet. They successfully implemented and tested their system with COTS hardware.

3.2.4.5. Analysis of Location and Routing Data for Attack Detection

In the following the last part of our review of related work in the field of cross data analysis of location related data is presented. These are mechanisms for consistency checks of routing protocol data reflecting the network topology and location data representing the position of a node within a specific topography.

[HPJ03] presents a general mechanism called packet leashes for detecting and thus defending against wormhole attacks. The notion leash means additional information is appended to the data messages. They differentiate between two sorts of leashes:

- *Geographical leashes* serve to limit the distance between two nodes. The sender attaches a time stamp and his location to a packet, the clock time of the nodes must be approximately synchronized. The recipient can compare his time and location information after the reception and compare it with that of the packet and decide whether these values are in a plausible proportion to each other.
- *Temporal leashes* serve to limit the network propagation time of a packet. The sender attaches an encoded and very precise time stamp to all packets; this requires very exactly synchronized clocks on the network nodes. So the recipient can decide again whether a packet had taken an unrealistically short time to get to him. The sender can additionally define a validity time for each packet.

This can be used both in (pro-)active protocols such as OLSR and reactive methods such as AODV, because they do not directly influence the routing but are based upon an extension of the data messages. It is mentioned in the article that geographic leashes are advantageous because they can be used in conjunction with a radio propagation model (e. g., allowing the detection of tunnels through obstacles), however this idea is not investigated further.

In [WB04] a mechanism to detect wormholes attacks, called MDS-VOW, is proposed which is based on the reconstruction of the sensors' layout using multi-dimensional scaling and a surface smoothing scheme. Wormholes are detected by visualizing the anomalies introduced by the attack caused by fake connections through the wormhole. This approach also tries to use node positions and network connectivity for plausibility checks and to detect attacks on a global scale. The proposed system was developed for sensor networks with stationary nodes and therefore is not directly applicable for MANETs.

[LPM*05] presents a solution that utilizes a combination of location information and cryptographic mechanisms to prevent a wormhole attack. Geometric random graphs induced by the communication range constraint of the nodes are used to derive conditions for detecting and defending against wormholes. Some communication range boundary limits are assumed, however, the actual signal strength is not used to estimate node distances. Other approaches try to detect anomalies of the network topology, e. g., suspicious changes in the routing table e. g., [TC05].

[Wan06] proposes a mechanism to detect wormhole attacks utilizing node positions and radio signal propagation time. An originator node retrieves the position of a destination node during route discovery. Then it estimates a lower bound of hops for the route based on radio signal propagation time and the distance to the destination and checks if the selected route is at least as long as the estimated shortest path length. However, no evaluation of the proposed mechanisms is presented and the applicability of signal propagation time measurements in real networks is not discussed.

Detection mechanisms for VANET nodes that are cheating about their location are proposed in [LSK06, LMSK06, LSKM10] to prevent attacks on geographic routing. The presented solution does not rely on special hardware or dedicated infrastructure.

3.3. Cooperative Trust Assessment

The trustworthiness of information sources and the assigned confidence level of available information have major impact on situation assessment. Potential threats to the SA system, e. g., injection of false data, should be detected and the trustworthiness of other nodes efficiently and cooperatively assessed.

Govindan and Mohaptra [GM12] compiled a detailed survey about trust computations and trust dynamics in MANETs. They compare proposed approaches and analyze prediction and aggregation algorithms as well as the impact of trust on security services.

We present in this section related work about cooperative trust assessment. We start by giving an overview on terms and definitions related to trust modeling and describe mechanisms for trust assessment. Afterwards we describe related work for cooperative trust assessment in distributed environments.

3.3.1. Trust Modeling

Many definitions for trust and reputation have been proposed in the literature but there is still some ambiguity and confusion how these terms should be used. We start by giving more abstract view on the concept trust and then present some more technical definitions.

Sztompka [Szt00] describes three types of orientations of human beings toward the future.

- *Hope* (opposite: resignation) is *"a passive, vague, not rationally justified feeling that things will turn out to the good (or to the bad)"*.
- *Confidence* is *"a still passive, but more focused and to some extent justified, faith that something good will happen (or not)"*.
- *Trust* is *"a bet about the future contingent actions of others"*.

He points out that the third type of orientation is more *"actively anticipating and facing an unknown future"*.

In the following we present some more technical definitions of trust and reputation that were proposed in the literature:

- Gambetta [Gam88] defines *trust* as *"a particular level of the subjective probability with which an agent assesses that another agent or group of agents will perform a particular action, both before he can monitor such action (or independently of his capacity ever to be able to monitor it) and in a context in which it affects his own action"*. This is similar to the definition of Sztompka trust as it also defines trust as a probabilistic variable.
- Abdul-Rahman and Hailes [ARH00] present a trust model and reputation mechanisms that are based on real-world social trust characteristics. They define *reputation* as *"an expectation about an agent's behavior based on information about or observations of its past behavior"*.
- Mui *et al.* [MMH02] point out the following relationships between the concepts of reciprocity, trust and reputation for agents embedded in social network:
  - Increase of an agent's reputation should increase the trust from the other agents.
  - Increase in an agent's trust should increase the likelihood that the agent will positively reciprocate to another agent's action.
  - Increase in reciprocating actions to other agents should increase the agent's reputation.

  A decrease of one of these values should lead to inverse effects. They define the three concept in the following way:
  - *Reciprocity* is the *"mutual exchange of deeds (such as favor or revenge)"*.
  - *Reputation* is the *"perception that an agent creates through past actions about its intentions and norms"*.
  - *Trust* is the *"subjective expectation an agent has about another's future behavior based on the history of their encounters"*.
- Wang and Vassileva [WV03] define *trust* as *"a peer's belief in another peer's capabilities, honesty and reliability based on its own direct experiences". They define reputation as "a peer's belief in another peer's capabilities, honesty and reliability based on recommendations received from other peers"*.

- Hussain and Chang [HC07] present an overview of definitions of trust and reputation in the literature and discuss their shortcomings. They particularly point out that some of the definitions of trust fail in respect to the following aspects:
  - *context dependent* nature of trust,
  - *time-dependent and dynamic* nature of trust, and that
  - trust is based on the *willingness and capability* of entities to act in a mutually agreed manner.

There are also some approaches that complement the trust assessment of an entity with a measure about the level of certainty of such a statement. Modeling trust not only as a single value enables particularly in dynamic, distributed environments more efficient mechanisms cooperative trust assessment.

Theodorakopoulos and Baras [TB04] propose to separate the opinion about the trustworthiness of another node and the level of certainty of this statement. They divide trustworthiness information into two independent elements: *trust* and *confidence* and give the following definitions:

Definition 3.3. *Trust is the "issuer's estimate of the target's trustworthiness."*

Definition 3.4. *Confidence is the "accuracy of the trust value assignment."*

Zouridaki *et al.* [ZMHT05] propose to map trust and confidence to a single value to simplify usage within applications. They combine trust and confidence metrics to a value that they call *trustworthiness*. This is a general opinion metric which can be tuned for specific application requirements based on two parameters and which can easily be incorporated into network decisions, e. g., route selection. They give the following definition:

Definition 3.5. Trustworthiness $T(t,c)$ *for specific trust and confidence values* $t, c \in \mathbb{R}$ *is defined by the following equation:*

$$T(t,c) = 1 - \frac{\sqrt{\frac{(t-1)^2}{x^2} + \frac{(c-1)^2}{y^2}}}{\sqrt{\frac{1}{x^2} + \frac{1}{y^2}}} \tag{3.7}$$

The intention of this concept is to consider trust values t whose corresponding confidence value c is high as more relevant to the calculation. Alternatively the smaller the confidence value c, the smaller the likelihood that the trust value t can be considered as valid. The result is that confidence gets larger as more trust values are added to the weighting and smaller as fewer trust values are added in the calculation of $T(t,c)$. The parameters x and y in the equation above can be used to adjust the influence of trust and confidence on the trustworthiness to the needs of a particular application.

Li and Wu [LW07] criticize that many existing reputation estimation approaches distinguish between right and wrong but ignore another very important aspect: *uncertainty*. Therefore they propose within their Mobility Assisted Uncertainty Reduction Scheme (MAURS) to divide reputation into three parts (uncertainty, belief, disbelief). The uncertainty metric reflects a node's confidence in the sufficiency of its evidence. Additionally they propose to exploit node mobility, a core characteristic of MANETs, as it increases the probability that two nodes get into direct contact. The proposed system reduces the overall uncertainty in the network and in particular the uncertainty in far-flung nodes.

3.3.2. Trust Assessment

In the following we present some approaches, i. e. how can we actually calculate trust values of entities based on past and current observations. Many reputation systems that were proposed in the literature are intuitive and ad hoc but beta distribution is grounded on a firm basis in the theory of statistics [CJI02]. It is a good prior density if a property is binomially distributed, i. e., successes and failures occur independently.

The trustworthiness of other nodes can be estimated based on local observation of their behavior and the result of these observations can be aggregated and correlated with probabilistic Bayesian analysis using beta distribution to determine reputation values.

Bayes' theorem shows how to calculate the probability of a belief (v) given an observation A:

$$P(B_i|A) = \frac{P(A|B_i)P(B_i)}{\sum_{i=1}^{N} P(A|B_i)P(B_i)}. \tag{3.8}$$

The prior distribution reflects the initial belief which may reflect also ignorance or indifference towards the initial situation.

Buchegger and Le Boudec [BLB03, Buc04] propose a distributed reputation system based on a Bayesian approach as follows. Node i models the behavior of other nodes j, which is assumed to misbehave with probability θ and the outcome is drawn independently from observation to observation. The true probability of a node to act maliciously is unknown and every node i may believe in different parameter values for θ. The estimation of θ is calculated using data obtained by direct or indirect observations based on the Bayes' theorem.

They propose to update the reputation values at each observation to integrate newly available knowledge and in order to increase the accuracy. The Beta function is the conjugate prior for binomial likelihood and thus the posterior density is also Beta:

$$f(\theta) = Beta(\alpha, \beta) = \frac{\Gamma(\alpha+\beta)}{\Gamma(\alpha)\Gamma(\beta)}\theta^{\alpha-1}(1-\theta)^{\beta-1}$$

$$\Gamma(x+1) = x\Gamma(x), \qquad \Gamma(1) = 1$$

Reputation values are updated in the following way:

- The prior distribution is $Beta(\alpha, \beta)$. It is initially it is set to $Beta(1,1)$ which represents a uniform distribution on [0,1] and means the absence of information.
- Secondly the posterior distribution is calculated and every new observations, e. g., s *observed correct behaviors* and f *observed misbehaviors*, are available it is updated according to the following equation:

$$\alpha := \alpha + s \tag{3.9}$$

$$\beta := \beta + f \tag{3.10}$$

These two parameters that are continuously updated reflect the current reputation values. The more evidence samples have been taken, the higher and narrower the peak of the beta distribution. The moments of Beta function (expectation value, variance) can be calculated in the following manner [BLB03, CJI02]:

$$E(Beta(\alpha, \beta)) = \frac{\alpha}{\alpha+\beta} \tag{3.11}$$

$$\sigma^2(Beta(\alpha,\beta)) = \frac{\alpha\beta}{(\alpha+\beta)^2(\alpha+\beta+1)} \tag{3.12}$$

A different approach was proposed by Xiong and Liu [XL02] for peer-to-peer networks. They claim that trust models are more accurate and effective if they do not only incorporate feedbacks from other peers. They propose to also incorporate additional information in the evaluation of the trustworthiness of peers such as the total number of transactions and the credibility of the feedback sources. They introduce three important trust parameters for a peer: the amount of satisfaction obtained from other peers through interactions, the total number of interactions with other peers and a balancing factor of trust for feedback source. The balancing factor is used to offset the potential of false feedback of peers

Aging Mechanisms Aging mechanisms are used to gives less weight to evidence received in the past to allow for reputation fading. Buchegger and Le Boudec [Buc04] propose a *modified Bayesian approach* that includes an aging mechanism. For this purpose they introduce a moving weighted average based on the individual observations made by node i for node j. If this observation is qualified as misbehavior s is set to 1 ($s = 1$) and $s = 0$ otherwise. The reputation value is updated according to the following equation

$$\alpha := u\alpha + s \tag{3.13}$$

$$\beta := u\beta + (1-s), \tag{3.14}$$

where u is a discount factor for past experiences. During inactivity periods: $\alpha := u\alpha$ and $\beta := u\beta$ values are periodically decayed.

Ganeriwal and Srivastava [GS04] propose a reputation-based framework for sensor networks as they argue that conventional security mechanisms based on cryptography alone are not sufficient. They propose that recently obtained information should be given more weight based on a suitable aging algorithm. For this purpose they also include a discount factor similar to the approach by Buchegger and Le Boudec described above.

An alternative aging approach is a *windowing mechanism*. Zouridaki *et al.* [ZMHT05] present a sliding window averaging mechanism in order to systematically expire old observation data. This way the accuracy and the fidelity of the opinion metric should be improved.

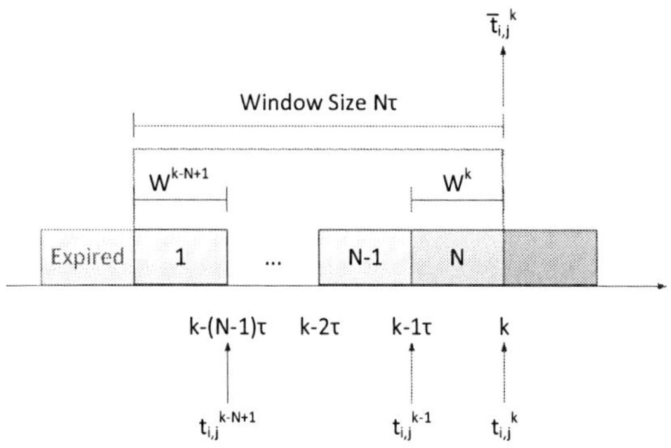

Figure 3.7.: Averaging and Observation Windows (based on [ZMHT05])

Observations are binned into observation of length τ, where the lth observation window is denoted as W^l (as shown in Figure 3.7). The general ideas is that for each observation window W^l the trust value $t_{l,j}^{k-N+l}$ of node i for node j is calculated. A sliding windowing averaging mechanism consisting of the N most recent observation windows is used to systematically expire old observation data. The averaged opinion value is calculated as a simple linear weighted average of all N opinion values computed during the sliding window (of N observation windows):

$$\bar{t}_{i,j}^{k} = \frac{2}{N(N+1)} \sum_{l=1}^{N} l\, t_{i,j}^{k-N+l} \tag{3.15}$$

3.3.3. Trust and Cooperation

In this section we describe related work for cooperative trust assessment in distributed environments. The assessment of the trustworthiness of information sources is a core component for situation assessment. Information collected from other nodes has to be authentic and protected, however even strong encryption and authentication methods can not perfectly guarantee confidentiality and integrity of these values.

A cooperative mechanism for measurement and evaluation of the trustworthiness of entities is required. A cooperative and distributed approach can counteract the damage a singular corruption of trustworthiness values may have. The general idea is that all MANET nodes locally run a trust assessment module which creates profiles for all other participating nodes in the network. The profiles store information about the reputation of every entity. The goal of this mechanism is to create trust information resulting from the reputation data which can be used to identify misbehavior nodes. This approach is based on a cooperative method to exchange information between the nodes. Data evaluation is performed locally on every node.

In [XL02] an approach for using trust values within peer-to-peer communication applications is defined. This principle is enhanced in [Buc04] by evaluating neighbors with a modified Bayesian approach and disseminating this information to other nodes. Mechanisms described in [GS04] are designed for sensor networks but could also be adapted for local evaluation of nodes and dissemination of reputation information within a MANET. In [ZMHT06] (based on [ZMHT05]) the combination of first hand reputation measurements (i. e. measurements performed directly by indicated nodes) with second hand reputation information is proposed in order to evaluate the most efficient route to a destination node.

In [GS04] a reputation-based framework for high integrity sensor networks is described. The main idea is a cooperative trust assessment system where each node collects information locally and information about directly adjacent nodes and stores the gathered information in a reputation table. Secondly the node collects reputation information from other network nodes and merges it into a global reputation table depending on the reputation of the source. The mechanisms are designed for sensor networks but may be adapted for local evaluation of nodes and dissemination of reputation information within a MANET.

Buchegger *et al.* [BMB08] discuss that reputation systems have been proposed and develop for diverse application areas. However there has been a duplication of efforts and still no consistent terminology is available. They outline common features and fundamental questions that are relevant in respect to reputation systems in order to bring together the different efforts. Certain principles must be considered in order to properly evaluate any self-organized reputation based system

In a cooperative trust assessment system all nodes locally run a trust assessment module that creates and updates trust profiles for all other participating nodes in the network. Trust profiles constructed in this way contain reputation information of every network node based on primary measurements given by the trust estimation module, e. g., retrieved from a local IDS, and secondary reputation information from neighbor nodes. The goal of this mechanism is to create a local database of trust information which can be used to identify misbehaving nodes. Different approaches that use a cooperative method to exchange information between nodes have been proposed for different fields.

Buchegger *et al.* [BMB08] discuss the concept of confirmation bias. Whereas direct observations should always be accepted, second-hand information should be checked and only information that does not differ too much from current expectations is acceptable. This concept of confirmation can also be motivated by observations in everyday life. Direct observations are regarded as undeniable facts and may be included despite the bias toward only accepting confirming information. Third-party indirect observations are subject to confirmation bias and is only accepted if it compatible with the user's view on reality. Additionally, they list as one of their lessons learned that deviation tests mitigate spurious ratings. Overall they argue for *"using deviation tests, discounting, passing on only first-hand information, introducing secondary response, and stressing the importance of identity"* [BMB08].

Cheng *et al.* [CGM11b]] propose a probabilistic, rendezvous based trust propagation model. Nodes that want to assess the trustworthiness of other nodes in the network (trust requesters) send out Trust Request (TR) tickets into the network. Nodes that have evaluated neighbor nodes (trust providers) may create Computed Trust (CT) tickets. The idea is that corresponding TR and CT tickets will meet with a certain probability at a common rendezvous node. When this happens the computed trust is propagated back to the trust requester. Their evaluation results show that their proposed method significantly reduces communication overhead compared to flooding based trust propagation models.

In another paper [CGM11a] Cheng *et al.* propose to exploit the mobility of MANET nodes for trust propagation. The approach is based on a mobility pattern estimation of neighbor nodes. Nodes that move in a similar direction are considered as less useful for trust propagations. The main idea is to base forwarding of trust information on movement estimation. Their evaluation results show that detection rates are significantly improved in comparison to static approaches. However, convergence time and communication overhead highly depend on the assumed mobility model

Trustworthiness Evaluation of Remote Nodes Beth *et al.* [BBK94] distinguish between two types of trust: *direct trust* and *recommendation trust.* Directly trusting in an entity means to believe in its capabilities whereas recommendation trust expresses the belief in an entity to be capable of deciding whether a third entity is reliable or not.

Maurer [Mau96] proposes an approach for evaluating trustworthiness of remote entities based on certificates. From a user point of view the question is: Which public keys should be considered authentic and which other users trustworthy? The assessment is based on a collection of certificates, e. g., retrieved from a PKI, and recommendations obtained from other users. An important aspect is the potential verification of several different certificate chains for the validation of a specific public key. A stronger recommendation with a higher confidence value may be achieved if several independent recommendations can be combined. For these purposes a measure for confidence is required. Maurer proposes a scale from 0 to 1, which represents no confidence at all to complete confidence. These values should be interpreted as probabilities of well-defined events of the same random experiment, which is non-trivial. A user has an initial set of statements S_A (certificates, recommendations and initial authenticity and trust assignments) that are considered true and new sets of statements are derived based on these statements and two inference rules (for authenticity of certified public-keys and trust). A statement $S \in S_A$ is not necessarily valid within the proposed probabilistic model but only with some probability. Based on these assumptions Maurer defines confidence as follows: *"The confidence value of a statement S, denoted $conf(S)$, is the probability that it can be derived from S_A."*

Abdul-Rahman and Hailes [ARH00] present a model where trust is composed of direct trust and recommender trust. This way a combined reputation value is derived for a subject in question based on personal opinions and the opinions of others. They point out that agents sometimes have to make trust decisions without being able to evaluate all aspects of a given situation. Therefore they must rely on propagating reputation mechanisms that provide information from other sources by means of word-of-mouth. They illustrate this by the following example *"a dishonest grocery store (owner) will quickly gain a reputation for dishonesty in the surrounding neighborhood and will in the long run be forced to close shop or improve its reputation"*.

In the computational model presented by Mui *et al.* [MMH02] trust expresses the expectation of one agent about the intention for reciprocation of another agent. The dyadic reciprocity between agents is measured as the proportion of cooperation actions in respect to the total number of encounters between two agents.

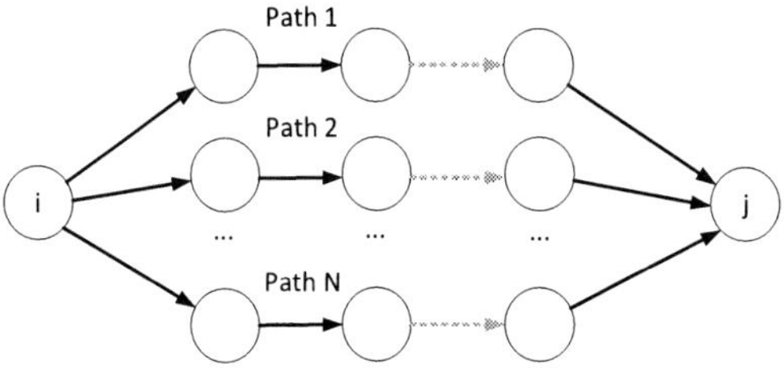

Figure 3.8.: Trust Propagation in Parallel Network via P Paths (based on [MMH02])

Mui *et al.* [MMH02] consider propagation mechanism for reputation via several paths in parallel. In a parallel network of P paths between two agents i and j (as shown in figure 3.8) an agent i needs to combine the evidence about j. For this purpose a measure of the reliability is required to weight all the evidences available via P different paths. They propose to use the number of encounters between agents as reliability measure. The weight increases with the number of encounters and above a specific level of sample size the estimator is expected to be reliable.

The weight of a complete chain is calculated in a multiplicative way based on the individual values in order to account for a break by an unreliable link on the path. The overall estimate across all paths is calculated as weighted sum across all paths. The proposed reputation propagation mechanisms, however, apply only to parallel networks.

Josang and Ismail [CJI02] propose a reputation mechanisms that incorporates observations provided by other nodes. In this case the information from highly reputed network nodes carries more weight than the feedback from other nodes with low reputation rating. They propose two mechanisms for *evaluating reputation values in node chains*, belief discounting and reputation discounting, in order to integrate reputation information based on the reputation of the agent who provided the feedback.

Kamvar *et al.* [KSGM03] present EigenTrust, a distributed algorithm for peer-to-peer networks to compute global trust values. They propose a system that aggregates local trust values of other users "in a natural manner". The approach is based on the idea of transitive trust, i. e., a peer i has a high opinion of those peers that acted benignly in previous transactions. The opinions of those peers is therefore considered as trustworthy based on the assumption that peers that are honest in file sharing transactions are also likely to be honest in respect to the local trust values that they provide.

The local normalized trust value of peer i for peer j is denoted by c_{ij} is calculated based on local observations. Peer i iteratively calculates the global trust value t_{ij} of a peer j based on the following equation:

$$t_{ij}^{(n+1)} = \sum_k c_{ik} t_{kj}^{(n)}, \qquad \text{where} t_{kj}^{(0)} = c_{kj}. \tag{3.16}$$

Buchegger and Le Boudec [BLB03] propose mechanisms for reputation updates and integration based on a Bayesian approach. First-hand observations are locally exchanged between neighboring nodes but also reports from distant nodes are integrated, which they call rumors. For example, when node i receives evidence from node j about a third node k she adjusts her opinion for j based on the deviation between her current opinion about node k and the evidence provided by j. If this deviation is bigger than a specific threshold the newly provided evidence is rejected and node i lowers her opinion about node j. If the deviation is small enough node i integrates the new evidence with her existing opinion about node k and increases her trust value for node j. They argue that the detection of malicious nodes in MANETs can significantly be sped up by utilizing second-hand information.

Buchegger and Le Boudec [Buc04] propose a distributed reputation system for peer-to-peer networks where first-hand reputation information is exchanged with other peers and second-hand reputation information is only accepted if it is compatible with current reputation ratings. Only first-hand information is exchanged, no reputation or trust ratings.

They define a reputation rating as an opinion about another node's behavior in the peer-to-peer network which is updated when first-hand observation is gathered or a reputation rating published by some other node is accepted and copied. When new second-hand information arrives a deviation test is performed and if the deviation is above a certain threshold the information is considered incompatible and not used. Otherwise the observations are considered and reputation ratings are adjusted accordingly.

Trust rating is an opinion about how honest a target node is within the reputation system which is updated whenever a node receives a report by some other node on first-hand information about this specific target node. The deviation test is always performed and trust values are updated accordingly; if the deviation test is positive the reputation ratings may be updated as well.

Theodorakopoulos and Baras [TB04] describe the trust evaluation process in MANETs as a path problem on a directed graph. The vertices of this graph represent MANET nodes and edges trust relations. They propose a reputation system based on semirings to evaluate a path to a distant node. Some evidence may be available for nodes that have met in the past but are not neighbors anymore. Confidence values for this information decreases over time and render it negligible after some time. Therefore the proposed system enables also establishing indirect trust relation between distant nodes that did not previously interact.

For the calculation of the opinion about a non-neighbor target node the individual trustworthiness values on the path to this target destination node are simply multiplied. If one of the links on the path has an associated trustworthiness value less than the default value we may infer that this nodes untrustworthy and hence we can consider the trustworthiness values of all nodes further down this path are as irrelevant.

They define two operators that are used to combine opinions: along a path and the across paths. Based on these operations a semiring with a partial order relation is defined. Two kinds of trust inference problems are discussed:

- *Finding the most trustworthy path:* The objective is to find the most trusted path from a node A to a destination node B, i. e., the path with the highest aggregated trust value. In this case: both trust and confidence values decrease when opinions are aggregated along a path. When opinions are aggregated across paths the opinion with the highest confidence is selected.
- Estimation of *(trust,confidence) values:* The objective of a node A is to assess and estimate the trustworthiness of a destination node B based on the (trust,confidence) values of intermediate nodes. In this case the (trust, confidence) pair $(t;c)$ is mapped to the so-called weight $(c/t;c)$. Opinions (represented by their weights) are combined along a path in the same way as parallel resistors: two resistors in parallel arrangement provide lower electric resistance than either of them by itself. When aggregating across paths the total trust value is the weighted harmonic average of the component opinions weighted proportional to their confidence values.

Zouridaki *et al.* [ZMHT06] extend their concept of trustworthiness to the notion of an opinion about the behavior of a target node by combining first-hand and second-hand trust information. A drawback of the old Hermes scheme [ZMHT05] is the lack of robustness with respect to malicious nodes that propagate erroneous trust information in the network. The newly proposed trust establishment scheme combines information that was obtained independently of other nodes (first-hand trust information) and recommendations (second-hand trust information) obtained via from other nodes.

3.4. Probabilistic State Modeling

Measurements and data sources in tactical MANETs are error-prone and vulnerable to attacks. Flexible and generally applicable system state modeling and assessment mechanisms are required as a basis for robust SA.

Therefore, we present in this section related work in the area of probabilistic state modeling. We start with the notation and terminology used for probabilistic state modeling and estimation problem. Afterwards we present a short overview of the technical background of particle filters (also known as Sequential Monte Carlo methods). Finally, we present state estimation approaches based on particle filters and describe in more details the challenges and proposed solutions in this area in respect to distributed state estimation and the incorporation of additional information sources.

3.4.1. State Modeling and Estimation

In this section we give a general introduction and mathematical specification of the probabilistic state estimation problem which is considered as a core building block for SA.

3.4.1.1. Notation and Terminology

Table 3.1 shows the definitions of the relevant symbols. In the following we present the general stochastic model for state evolution and the objective of the state estimation process.

Assumptions The target state evolves according to a discrete-time stochastic model defined by the following equation:

$$\mathbf{x}_k = \mathbf{f}_{k-1}(\mathbf{x}_{k-1}, \mathbf{v}_{k-1}) \tag{3.17}$$

where $\mathbf{f}_{k-1} : \mathbb{R}^{n_x} \times \mathbb{R}^{n_v} \rightarrow \mathbb{R}^{n_x}$ is a known, possibly nonlinear function of the state $\mathbf{x}_{k-1}$ and $\mathbf{v}_{k-1}$ and $\mathbf{v}_k \in \mathbb{R}^{n_v}$ a process noise sequence.

Table 3.1.: Definitions of Relevant Symbols for State Estimation

Symbol	Description
$\mathbf{x}_k \in \mathbb{R}^{n_x}$	State vector (including continuous and discrete states)
$\mathbf{x}_{c,k} \in \mathbb{R}^{n_{x_c}}$	Vector of continuous states
$\mathbf{x}_{d,k} \in \mathbb{R}^{n_{x_d}}$	Vector of discrete states (also called mode, regime or model)
$\mathbf{v}_k \in \mathbb{R}^{n_v}$	Process noise sequence
$\mathbf{z}_k \in \mathbb{R}^{n_z}$	Measurements related to target state
$\mathbf{w}_k \in \mathbb{R}^{n_w}$	Measurements noise sequence
$\mathbf{X}_k \triangleq \mathbf{x}_{1:k} = \{\mathbf{x}_i, i = 1, \ldots, k\}$	Sequence of all target states up to time k
$\mathbf{Z}_k \triangleq \mathbf{z}_{1:k} = \{\mathbf{z}_i, i = 1, \ldots, k\}$	Sequence of all available measurements up to time k
$\mathbf{z}_0$	set of no measurements
$\mathbf{f}_k : \mathbb{R}^{n_x} \times \mathbb{R}^{n_v} \rightarrow \mathbb{R}^{n_x}$	Known, possibly nonlinear function of the state $\mathbf{x}_{k-1}$ and $\mathbf{v}_{k-1}$
$\mathbf{h}_k : \mathbb{R}^{n_x} \times \mathbb{R}^{n_w} \rightarrow \mathbb{R}^{n_z}$	Known, possibly nonlinear function of the state $\mathbf{x}_k$ and the measurement noice sequence $\mathbf{w}_k$
$p(\mathbf{x}_0) \triangleq p(\{\mathbf{x}_0 \| \mathbf{z}_0)$	Probability density function (PDF) of initial target state
$p(\mathbf{x}_k \| \mathbf{Z}_k)$	Posterior PDF of target state
$n_x \in \mathbb{N}$	Dimension of state vector $\mathbf{x}$
$n_{x_c} \in \mathbb{N}$	Dimension of vector of continuous states $\mathbf{x_c}$
$n_{x_d} \in \mathbb{N}$	Dimension of vector of discrete states $\mathbf{x_d}$
$n_v \in \mathbb{N}$	Dimension of process noise sequence vector $\mathbf{v}$
$n_z \in \mathbb{N}$	Dimension of measurement vector $\mathbf{z}$
$n_w \in \mathbb{N}$	Dimension of measurement noise sequence vector $\mathbf{w}$
$k \in \mathbb{N}$	Time index assigned to a continuous-time instant t_k
$\mathbb{N}$	Set of real numbers
$\mathbb{R}$	Set of natural numbers

Noise sequences $\mathbf{v}_{k-1}$ and $\mathbf{w}_k$ are independent and identically-distributed process noise sequences: white with known probability density functions and mutually independent. The probability density function of the initial target state $p(\mathbf{x}_0)$ is independent of noise sequences.

Objective The overall objective of our architecture is to recursively estimate $\mathbf{x}_k$ based on measurements $\mathbf{z}_k \in \mathbb{R}^{n_x}$ that are related to the target state via the measurement equation:

$$\mathbf{z}_k = \mathbf{h}_k(\mathbf{x}_k, \mathbf{w}_k) \tag{3.18}$$

where $\mathbf{h}_k : \mathbb{R}^{n_x} \times \mathbb{R}^{n_w} \to \mathbb{R}^{n_z}$ is a known, possibly nonlinear function of the state $\mathbf{x}_k$ and the measurement noise sequence $\mathbf{w}_k$. $\mathbf{w}_k \in \mathbb{R}^{n_w}$ is the measurements noise sequence.

3.4.1.2. Hybrid State Estimation

In many application scenarios in tactical environments the system state also includes a set of discrete state variables. For example, discrete state variables are used to characterize a mode of operation, a behavioral pattern or an object type in tracking systems.

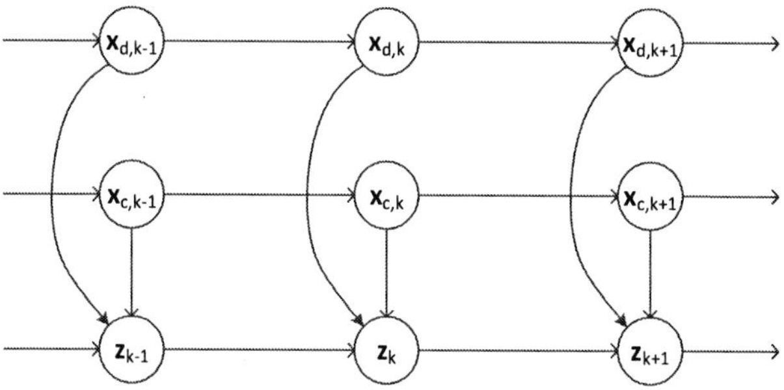

Figure 3.9.: Hybrid State Space Model [DLNP05, Kaw12]

In this case the state vector is an augmented hybrid state vector $\mathbf{x}_k$ comprising continuous states $\mathbf{x}_{c,k}$ as well as discrete states $\mathbf{x}_{d,k}$ (see figure 3.9). The discrete state vector $\mathbf{x}_{d,k}$ represents a specific element of a set $S \triangleq \{1, 2, \ldots s\}$ of modes (regimes).

In the following we describe the model for hybrid state estimation in more details similar to the stochastic model described above. The hybrid system model can be described a using the following target state evolution and measurement equations [DWS*04, RAG04]:

$$\mathbf{x}_{c,k} = \mathbf{f}_{k-1}(\mathbf{x}_{c,k-1}, \mathbf{x}_{d,k}, \mathbf{v}_{k-1}) \tag{3.19}$$

$$\mathbf{z}_k = \mathbf{h}_k(\mathbf{x}_{c,k}, \mathbf{x}_{d,k}, \mathbf{w}_k) \tag{3.20}$$

where $\mathbf{x}_{d,k}$ is the vector of discrete states representing the mode (regime) in effect during time period $(t_{k-1}, t_k]$. The new mode $\mathbf{x}_{d,k+1}$ takes effect at time t_k.

The mode variable $\mathbf{x}_{d,k}$ is typically modeled as a first-order Markov chain with the following transitional probabilities:

$$\pi_{ij} \triangleq P\{\mathbf{x}_{d,k} = j | \mathbf{x}_{d,k-1} = i\} \qquad (i, j \in S) \tag{3.21}$$

where $\pi_{ij} \geq 0$ and $\sum_{j=1}^{s} \pi_{ij} = 1$ for each $i, j \in S$..

3.4.1.3. Kalman Filters

The Kalman filter assumes a linear dynamical system where all error terms and measurements have a Gaussian distribution. It was first derived by Kalman [Kal60] and Kalman and Bucy [KB61]. The target state evolution and the measurement equations are given by the following equations [RAG04]:

$$\mathbf{x}_k = \mathbf{F}_{k-1}\mathbf{x}_{k-1} + \mathbf{v}_{k-1} \tag{3.22}$$

$$\mathbf{z}_k = \mathbf{H}_k\mathbf{x}_k + \mathbf{w}_k \tag{3.23}$$

where $\mathbf{F}_{k-1}$ is a known matrix of dimensions $n_x \times n_x$ and $\mathbf{H}_{k-1}$ is a known matrix of dimensions $n_z \times n_x$. It is assumed that random sequences $\mathbf{v}_{k-1}$ and $\mathbf{w}_k$ are mutually independent white Gaussian distributed with zero mean value and covariances $\mathbf{Q}_{k-1}$ and $\mathbf{R}_k$ respectively.

Kalman filters are extremely efficient but not applicable for nonlinear state estimation problems. In this case the extended Kalmanfilter (EKF) and the unscenred Kalmanfilter (UKF) may provide good approximations. In general, the UKF is dominating the EKF [DWS*04].

Extended Kalman Filter The extended Kalman filter (EKF) is applied for nonlinear systems with additive noise. It can approximate also nonlinear state transition and observation models. The following target state evolution and measurement equations are assumed:

$$\mathbf{x}_k = \mathbf{f}_{k-1}(\mathbf{x}_{k-1}) +, \mathbf{v}_{k-1} \tag{3.24}$$

$$\mathbf{z}_k = \mathbf{h}_k(\mathbf{x}_k) + \mathbf{w}_k \tag{3.25}$$

where $\mathbf{v}_{k-1}$ and $\mathbf{w}_k$ are independent and identically-distributed process noise sequences: white with known probability density functions and mutually independent. As above the functions $\mathbf{f}_{k-1}$ and $\mathbf{h}_k$ are used to compute the predicted state and measurement.

The EKF is an analytic approximation as the Jacobians $\hat{\mathbf{F}}_{k-1}$ and $\hat{\mathbf{H}}_k$ have to be derived analytically. The resulting matrices are used in Kalman filter equations. The EKF basically linearizes the non-linear function based on an estimate of the current mean and covariance.

Unscented Kalman Filter The unscented Kalman filter (UKF) is a nonlinear Kalman filter that is used within the EKF framework but does not approximate the nonlinear functions $\mathbf{f}_{k-1}$ and $\mathbf{h}_k$. Instead, the posterior probability density $p(\mathbf{x}_k|\mathbf{Z}_k)$ is approximated using a deterministic sampling of points which represent the underlying distribution as a Gaussian density [RAG04].

Kalman filters are extremely efficient but not directly applicable for non-linear transformations. Approximations may be achieved using the extended Kalman filter (EKF) or the unscented Kalman filter (UKF) as described above.

In contrast, particle filters are not constrained by linear dynamics, linear measurements or Gaussian PDFs Particle filtering. They can also be applied to estimation problems with external constraints (e. g., road-constrained networks) or hybrid estimation problems comprising both discrete and continuous state variables. We give a more detailed introduction to particle filters in the following section.

3.4.2. Particle Fiter

In this section we outline the mathematical background of state estimation using particle filters and how they can be applied to our scenario. Particle filters do not require linearization of dynamic equations or measurement equations. Values of the measurement likelihood function are evaluated at discrete points in the state space [AS03]. Particle filter can also handle constrained estimation problems. Utilizing constraints information may lead to significantly better estimation results. Particle filters are able to handles hybrid state vectors including discrete state variables.

A particle filters are a sampling-based approximate inference algorithm capable of dealing with hybrid dynamic belief networks. (DBNs). From a Bayesian perspective the objective of a particle filter is to recursively quantify some degree of belief in the state $\mathbf{x}_k$ at time k given the data $\mathbf{z}_{1:k}$ up to time k. Thus it is required to construct the posterior PDF $p(\mathbf{x}_k|\mathbf{z}_{1:k})$ which can be obtained recursively in two stages: prediction and update. Data fusion techniques are used for state estimation (e. g., next expected routing message in a protocol stream, frequencies of events, route lengths, etc.). At a higher level the goal is to predict how the system is going to evolve and what other nodes will do in the near future.

Explicit knowledge should be modeled as probability density functions to be included in state estimation process and measurements and partial results should be exchanged for cooperative state estimation.

In the following a short overview of the basic concept of particle filters including importance (re-)sampling is given [DFG01, RAG04].

The posterior distribution $p(\mathbf{x}_{0:k}|\mathbf{z}_{1:k})$ and its associated features (including the marginal distribution $p(\mathbf{x}_k|\mathbf{z}_{1:k})$, known as filtering distribution) are recursively in time estimated as well as the expectations

$$I(g) = \mathbb{E}_{p(\mathbf{x}_k|\mathbf{Z}_k)}[g(\mathbf{x}_k)] \triangleq \int g(\mathbf{x}_k) p(\mathbf{x}_k|\mathbf{Z}_k)\, d\mathbf{x}_k \tag{3.26}$$

for some function of interest $g : \mathbb{R}^{n_x} \to \mathbb{R}^{n_g}$ integrable with respect to $p(\mathbf{x}_k$.

An example is the conditional mean, in which case $g_(\mathbf{x}_k) = \mathbf{x}_k$, and we get the minimum mean-square error (MMSE) estimate:

$$\hat{\mathbf{x}}_{k|k}^{MMSE} \triangleq \mathbb{E}\{\mathbf{x}_k|\mathbf{Z}_k\} = \int \mathbf{x}_k \cdot p(\mathbf{x}_k|\mathbf{Z}_k)\, d\mathbf{x}_k, \tag{3.27}$$

where the maximum a posteriori (MAP) estimate is the maximum of $p(\mathbf{x}_k|\mathbf{Z}_k)$:

$$\hat{\mathbf{x}}_{k|k}^{MAP} \triangleq \arg\max_{\mathbf{x}_k} p(\mathbf{x}_k|\mathbf{Z}_k). \tag{3.28}$$

Chapman-Kolmogorov equation The Chapman-Kolmogorov equation can be used to obtain the *prediction* PDF the state at time k:

$$p(\mathbf{x}_k|\mathbf{Z}_{k-1}) = \int p(\mathbf{x}_k|\mathbf{x}_{k-1})p(\mathbf{x}_{k-1}|\mathbf{Z}_{k-1})\,d\mathbf{x}_{k-1} \tag{3.29}$$

Note: The Chapman-Kolmogorov equation can be applied since equation (3.17) describes a Markov process of order one and therefore $p(\mathbf{x}_k|\mathbf{x}_{k-1},\mathbf{Z}_{k-1}) = p(\mathbf{x}_k|\mathbf{x}_{k-1})$.

The *update* stage involves an update of the prediction PDF via the Bayes' rule:

$$p(\mathbf{x}_k|\mathbf{Z}_k) = p(\mathbf{x}_k|\mathbf{z}_k,\mathbf{Z}_{k-1}) = \frac{p(\mathbf{z}_k|\mathbf{x}_k,\mathbf{Z}_{k-1})p(\mathbf{x}_k|\mathbf{Z}_{k-1})}{p(\mathbf{z}_k|\mathbf{Z}_{k-1})} = \frac{p(\mathbf{z}_k|\mathbf{x}_k)p(\mathbf{x}_k|\mathbf{Z}_{k-1})}{p(\mathbf{z}_k|\mathbf{Z}_{k-1})} \tag{3.30}$$

where the normalizing constant

$$p(\mathbf{z}_k|\mathbf{Z}_{k-1}) = \int p(\mathbf{z}_k|\mathbf{x}_k)p(\mathbf{x}_k|\mathbf{Z}_{k-1})\,d\mathbf{x}_k \tag{3.31}$$

depends on the likelihood function $p(\mathbf{z}_k|\mathbf{x}_k)$, defined by the measurement model (3.18) and the known statistics of $\mathbf{w}_k$.

The key concept of particle filters – also known as sequential Monte Carlo (SMC) methods – is to represent the probability density function by a set of samples (also referred to as particles) and their associated weights.

The the PDF $p(\mathbf{x}_k|\mathbf{z}_{1:l})$ is therefore approximated with an empirical density function:

$$p(\mathbf{x}_k|\mathbf{z}_{1:l}) \approx \sum_{i=1}^{N} \tilde{q}_k^i \delta(\mathbf{x}_k - \mathbf{x}_k^i), \qquad \sum_{i=1}^{N} \tilde{q}_k^i = 1, \qquad \tilde{q}_k^i \geq 0, \forall i. \tag{3.32}$$

where δ is the Dirac delta function and $\tilde{q}_k^i$ denotes the weight associated with particle $\mathbf{x}_k^i$.

3.4.2.1. Monte Carlo Integration and Importance Sampling

Monte Carlo Integration Monte Carlo (MC) integration is the basis of SMC methods. Its objective is the numerical evaluation of a multidimensional integral:

$$I(\mathbf{g}) = \int \mathbf{g}(\mathbf{x})\,d\mathbf{x} \tag{3.33}$$

where $\mathbf{x} \in \mathbb{R}^{n_x}$.

Monte Carlo methods factorize $\mathbf{g}(\mathbf{x}) = \mathbf{f}(\mathbf{x}) \cdot \pi(\mathbf{x})$ in such a way that $\pi(\mathbf{x})$ is interpreted as a probability density satisfying $\pi(\mathbf{x}) \geq 0$ and $\int \pi(\mathbf{x})\,d\mathbf{x} = 1$.

The assumptions is that it is possible to draw $N \gg 1$ samples $\{\mathbf{x}^i, i = 1, \dots, N\}$ distributed according to $\pi(\mathbf{x})$. An approximation of the integral

$$I(\mathbf{g}) = \int \mathbf{f}(\mathbf{x})\pi(\mathbf{x})\,d\mathbf{x} \tag{3.34}$$

we define the sample mean:

$$I_N(\mathbf{g}) = \frac{1}{N}\sum_{i=1}^{N}\mathbf{f}(\mathbf{x}^i). \tag{3.35}$$

If the samples $\mathbf{x}^i$ are independent then $I_N(\mathbf{g})$ is an unbiased estimate and according to the law of large numbers $I_N(\mathbf{g})$ will almost surely converge to $I(\mathbf{g})$. If the variance of $\mathbf{f}(\mathbf{x})$,

$$\sigma^2 = \int (\mathbf{f}(\mathbf{x}) - I)^2\pi(\mathbf{x})\,d\mathbf{x}$$

is finite, then the central limit theorem holds and the estimation error converges in distribution:

$$\lim_{N\to\infty} \sqrt{N}(I_N - I) \sim \mathcal{N}(0, \sigma^2).$$

Importance Sampling Ideally we want to generate samples directly from $\pi(\mathbf{x})$ and estimate I using (3.35). Suppose we can only generate samples from a density $q(\mathbf{x})$ which is similar to $\pi(\mathbf{x})$. Then a correct weighting of the samples makes the MC estimation still possible. The PDF $q(\mathbf{x})$ is referred to as the *importance* or *proposal density*

$$\pi(\mathbf{x}) > 0 \Rightarrow q(\mathbf{x}) > 0, \qquad \forall \mathbf{x} \in \mathbb{R}^{n_x} \tag{3.36}$$

which means that $q(\mathbf{x})$ and $\pi(\mathbf{x})$ have the same support.

Equation (3.34) can be rewritten as:

$$I(\mathbf{g}) = \int \mathbf{f}(\mathbf{x})\pi(\mathbf{x})\,d\mathbf{x} = \int \mathbf{f}(\mathbf{x})\frac{\pi(\mathbf{x})}{q(\mathbf{x})}q(\mathbf{x})\,d\mathbf{x} \tag{3.37}$$

provided that $\frac{\pi(\mathbf{x})}{q(\mathbf{x})}$ is upper bounded.

A Monte Carlo estimate of I can be computed by generating $N \gg 1$ independent samples $\{\mathbf{x}^i, i = 1, \ldots, N\}$ distributed according to $q(\mathbf{x})$ and forming the weighted sum:

$$I_N(\mathbf{g}) = \frac{1}{N}\sum_{i=1}^{N}\mathbf{f}(\mathbf{x}^i)\tilde{q}(\mathbf{x}^i), \qquad \text{where} \quad \tilde{q}(\mathbf{x}^i) = \frac{\pi(\mathbf{x}^i)}{q(\mathbf{x}^i)} \tag{3.38}$$

are the importance weights.

If the normalizing factor of the desired density $\pi(\mathbf{x})$ is unknown, we need to perform normalization of the importance weights. Then we estimate I_N as follows:

$$I_N = \frac{\frac{1}{N}\sum_{i=1}^{N}\mathbf{f}(\mathbf{x}^i)\tilde{w}(\mathbf{x}^i)}{\frac{1}{N}\sum_{i=1}^{N}\tilde{w}(\mathbf{x}^i)} = \sum_{i=1}^{n}\mathbf{f}(\mathbf{x}^i)w(\mathbf{x}^i) \tag{3.39}$$

where the normalized importance weights are given by:

$$w(\mathbf{x}^i) = \frac{\tilde{w}(\mathbf{x}^i)}{\sum_{j=1}^{N}\tilde{w}(\mathbf{x}^j)} \tag{3.40}$$

Sequential Importance Sampling Sequential importance sampling (SIS) is the application of importance sampling to perform nonlinear filtering. It is the basis for most sequential MC filters known as bootstrap filtering, condensation algorithm, particle filtering or survival of the fittest.

Let $\{\mathbf{X}_k^i, w_k^i\}_{i=1}^n$ denote a random measure that characterizes the joint posterior density $p(\mathbf{X}_k|\mathbf{Z}_k)$, where $\{\mathbf{X}_k^i, i = 1, \ldots, N\}$ is a set of support points with associated weights

$\{w_k^i, i = 1, \ldots, N\}$. The weights are normalized such that $\sum_i w_k^i = 1$. Then, the joint posterior density at k can be approximated as follows:

$$p(\mathbf{X}_k|\mathbf{Z}_k) \approx \sum_{i=1}^{N} w_k^i \delta(\mathbf{X}_k - \mathbf{X}_k^i). \tag{3.41}$$

If the samples $\mathbf{X}_k^i$ were drawn from an importance density $q(\mathbf{X}_k|\mathbf{Z}_k)$, then according to (3.38)

$$w_k^i \propto \frac{p(\mathbf{X}_k^i|\mathbf{Z}_k)}{q(\mathbf{X}_k^i|\mathbf{Z}_k)} \tag{3.42}$$

If $q(\mathbf{x}_k|\mathbf{X}_{k-1}, \mathbf{Z}_k) = q(\mathbf{x}_k|\mathbf{x}_{k-1}, \mathbf{z}_k)$, then the importance density becomes only dependent on the $\mathbf{x}_{k-1}$ and $\mathbf{z}_k$. In this case only $\mathbf{x}_k^i$ need be stored, and so one can discard the Path $\mathbf{X}_{k-1}^i$, and the history of observations, $\mathbf{Z}_{k-1}$.

In such scenarios the weight can be derived as

$$w_k^i \propto w_{k-1}^i \frac{p(\mathbf{z}_k|\mathbf{x}_k^i) p(\mathbf{x}_k^i|\mathbf{x}_{k-1}^i)}{q(\mathbf{x}_k^i|\mathbf{x}_{k-1}^i, \mathbf{z}_k)} \tag{3.43}$$

and the posterior filtered density $p(\mathbf{x}_k|\mathbf{Z}_k)$ can be approximated as

$$p(\mathbf{x}_k|\mathbf{Z}_k) \approx \sum_{i=1}^{N} w_k^i \delta(\mathbf{x}_k - \mathbf{x}_k^i). \tag{3.44}$$

It can be shown that as $N \rightarrow \infty$ the approximation approaches the true posterior density $p(\mathbf{x}_k|\mathbf{Z}_k)$.

Filtering via SIS thus consists of recursive propagation of importance weights w_k^i and support points $\mathbf{x}_k^i$ as each measurement is received sequentially.

3.4.2.2. Degeneracy Problem

Ideally the importance density function should be the posterior distribution itself. For an importance function as described above, it has been shown that the variance of importance weights can only increase over time. This leads to the degeneracy phenomenon: after a certain number of recursive steps, all but one particle will have negligible nor-

Table 3.2.: Filtering via Sequential Importance Sampling [RAG04]

$[\{\mathbf{x}_k^i, w_k^i\}_{i=1}^N] \leftarrow SIS[\{\mathbf{x}_{k-1}^i, w_{k-1}^i\}_{i=1}^N, \mathbf{z}_k]$

- FOR $i \leftarrow 1$ TO N
  - Draw $\mathbf{x}_k^i \sim q(\mathbf{x}_k | \mathbf{x}_{k-1}^i, \mathbf{z}_k)$
  - Evaluate the importance weights up to normalizing constant according to (3.43):

$$\tilde{w}_k^i \leftarrow w_{k-1}^i \frac{p(\mathbf{z}_k|\mathbf{x}_k^i) p(\mathbf{x}_k^i|\mathbf{x}_{k-1}^i)}{q(\mathbf{x}_k^i|\mathbf{x}_{k-1}^i, \mathbf{z}_k)}$$

- Calculate total weight: $t \leftarrow \sum_{i=1}^N \tilde{w}_k^i$.
- FOR $i \leftarrow 1$ TO N
  - Normalize: $w_k^i \leftarrow \frac{1}{t} \cdot \tilde{w}_k^i$

malized weights. A suitable measure of degeneracy of an algorithm is the effective sample size N_{eff} [RAG04]:

$$\hat{N}_{eff} = \frac{1}{\sum_{i=1}^N (w_k^i)^2}. \tag{3.45}$$

$1 \leq N_{eff} \leq N$ where $N_{eff} = N$ if the weights are uniform and $N_{eff} = 1$ if one weight is $w_k^j = 1$ and all other are 0 (maximum degeneracy).

Resampling Resampling is required whenever a significant degeneracy is observed, i. e.when $\hat{N}_{eff}$ falls below some threshold $\hat{N}_{thr}$. Resampling eliminates samples with low importance and multiplies samples with high importance weights. It involves a mapping of random measure $\{\mathbf{x}_k^i, w_k^i\}$ into a random measure $\{\mathbf{x}_k^{i*}, 1/N\}$ with uniform weights.

The new set of random samples $\{\mathbf{x}_k^{i*}, 1/N\}_{i=1}^N$ is generated by resampling (with replacement) N times from an approximate discrete representation of $p(\mathbf{x}_k|\mathbf{Z}_k)$ given by

$$p(\mathbf{x}_k|\mathbf{Z}_k) \approx \sum_{i=1}^N w_k^i \delta(\mathbf{x}_k - \mathbf{x}_k^i) \tag{3.46}$$

so that $P(\mathbf{x}_k^{i*} = \mathbf{x}_k^j) = w_k^j$.

A direct implementation of resampling would consist of generating N i.i.d. variables from the uniform distribution, sorting them in ascending order and comparing them with the cumulative sum of normalized weights. The complexity of this algorithm can be reduced by sampling N ordered uniforms using and algorithms based on order statistics.

Systematic resampling is an efficient scheme of computational complexity $\mathcal{O}(N)$ that minimizes the MC variation The pseudocode of this algorithm is described in table 3.3, where CSW stands for the cumulative sum of weights (CSW) of random measure $\{\mathbf{x}_k^i, w_k^i\}$. For each resampled particle x_k^{j*}, this resampling algorithm also stores the index of its parent denoted by i^j.

Table 3.3.: Resampling Algorithm [RAG04]

$[\{\mathbf{x}_k^{j*}, w_k^j, i^j\}_{j=1}^N] \leftarrow RESAMPLE[\{\mathbf{x}_k^i, w_k^i\}_{i=1}^N]$

- Initialize CSW: $c_1 \leftarrow w_k^1$
- FOR $i \leftarrow 2$ TO N
  - Construct CSW: $c_i \leftarrow c_{i-1} + w_k^i$
- Start at the bottom of the CSW: $i \leftarrow 1$
- Draw a starting point: $u_1 \sim \mathcal{U}[0, N^{-1}]$
- FOR $j \leftarrow 1$ TO N
  - Move along the CSW: $u_j \leftarrow u_1 + N^{-1}(j-1)$
  - WHILE $u_j > c_i$
    - * $i \leftarrow i+1$
  - Assign sample: $\mathbf{x}_k^{j*} \leftarrow \mathbf{x}_k^i$
  - Assign weight: $w_k^j \leftarrow N^{-1}$
  - Assign parent: $i_j \leftarrow i$

3.4.2.3. State Estimation

The objective is the determination of the posterior PDF

$$p(\mathbf{x}|\mathbf{z}_{1:k}^1, \ldots, \mathbf{z}_{1:k}^N)$$

As the following property holds [RGT03]

$$p(\mathbf{x}_{1:k}|\mathbf{z}^1_{1:k},\dots,\mathbf{z}^N_{1:k}) \quad \propto \quad p(\mathbf{x}_{1:k})\prod_{i=1}^{N}\frac{p(\mathbf{x}_{1:k}|\mathbf{z}^i_{1:k})}{p(\mathbf{x}_{1:k})} \tag{3.47}$$

the following representation of the posterior PDF can be derived:

$$\begin{aligned}
p(\mathbf{x}_{1:k}|\mathbf{z}^1_{1:k},\dots,\mathbf{z}^N_{1:k}) &\propto p(\mathbf{x}_{1:k})\; p(\mathbf{z}^1_{1:k},\dots,\mathbf{z}^N_{1:k}|\mathbf{x}_{1:k})\\
&= p(\mathbf{x}_{1:k})\prod_{i=1}^{N}p(\mathbf{z}^i_{1:k}|\mathbf{x}_{1:k})\\
&\propto p(\mathbf{x}_{1:k})\prod_{i=1}^{N}\frac{p(\mathbf{x}_{1:k}|\mathbf{z}^i_{1:k})}{p(\mathbf{x}_{1:k})}
\end{aligned}$$

A problem that needs to be considered — in particular for distributed and cooperative state estimation — is that the evidence (relevant measurements) $\frac{p(\mathbf{x}_{1:k}|\mathbf{z}^i_{1:k})}{p(\mathbf{x}_{1:k})}$ grows over time and cannot be exchanged in messages with fixed length. Furthermore measurements can only be combined if they refer to the same state; and therefore the same instance in time. Only simultaneously gathered observations can be directly combined.

3.4.3. State Estimation Approaches based on Particle Filter

In the following we give an overview on hybrid particle filter based approaches incorporating discrete state modeling. Then we present in more details the specific challenges of distributed state estimation and proposed solutions for the incorporation of additional information sources.

Schön [Sch06] addresses the problem of state and parameter estimation in nonlinear and non-Gaussian dynamic systems using particle filters. These proposed methods rely on models of the underlying system. The parameter estimation problem is addressed for a relatively general class of mixed linear/nonlinear state-space models. The proposed system is applied to and evaluated for a camera positioning problem arising from augmented reality and sensor fusion problems originating from automotive active safety systems.

3.4.3.1. Hybrid Approaches Including Discrete State Modeling

Hofbaur and Williams [HW02] present an approach for hybrid estimation of complex systems enabling to track and model a system dynamics in different discrete behavioral modes. They merge hidden Markov models (HMM) with continuous dynamical system models introducing the concept of concurrent probabilistic hybrid automata (cPHA). They further elaborate on their hybrid estimation framework (HME) in [HW04] and analyze shortcomings of multiple model estimation approaches. They develop a scalable scheme that is able to efficiently estimate hybrid states in complex systems with large number of modes. For this purpose they regard the estimation task as a search problem focusing on highly probable hypotheses neglecting unlikely hypotheses. Their evaluation results show similarly good estimation performance compared to multi model estimation but providing scalability up to systems of high complexity.

Funiak [Fun04] presents a hybrid approach for state estimation with particle filters which combines Rao-Blackwellised particle filtering with a Gaussian representation. Trajectories are traced by the discrete variables over time and for each trajectory the continuous state is estimated with a Kalman Filter.

They use position tracking of a robotic arm as application example. The position of the arm is only indirectly observed as the sensors measure the arm angle and the exerted force. The system dynamics changes when an additional weight, e. g., a ball, is present at the end of the arm. Therefore, they model the system mode using a discrete variable *has-ball* with two discrete modes (*true*, *false*) representing whether or not the robot carries a ball.

The continuous state of the robot is modeled using four variables: two describing the angles of the position of the robotic arm (ω_1 and ω_2) and two representing the torque exerted by the actuator at the center of the robot (θ_1 and θ_2). If the robot carries a ball (*has-ball* is *yes*) and $\theta_1 > 0.7$ the probability of transitioning to mode no and staying in mode yes are the same. If $\theta_1 < 0.7$ the robot will keep the ball with probability 1 (see figure 3.10). Similarly the system will stay in the other state (without ball) if $\theta_1 < 0.7$ and may change state otherwise.

Dearden *et al.* [DWS*04] propose a hybrid discrete/continuous subsystem model for Mars rovers enabling real-time fault detection and situational awareness. They propose an approach based on a particle filters as they are also applicable nonlinear models with

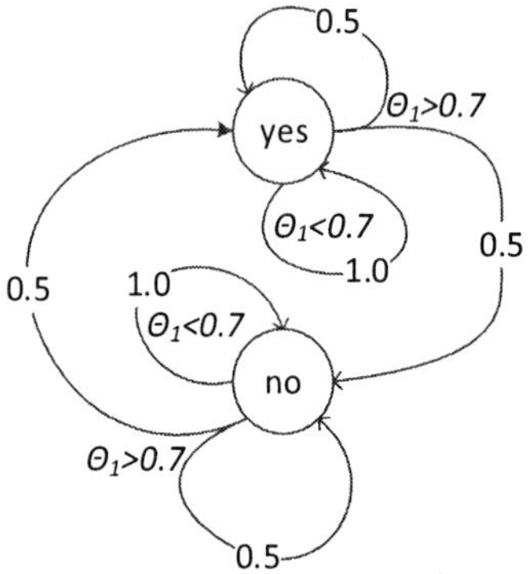

Figure 3.10.: Hybrid State Robot Example [Fun04]

arbitrary prior belief distributions. They acknowledge that there has been much work on Rao-Blackwellized Particle Filtering (RBPF) on combining PFs and KFs for tracking linear multi-modal systems with Gaussian noise. However, RBPF are restricted to linear problems with Gaussian noise which is not the case for many of the problems that they address. Their proposed Gaussian particle filter (GPF) allows propagating an approximation of the continuous state variables by updating a Gaussian approximation for each particle of the continuous parameters using an unscented Kalman filter.

Kawamoto [Kaw12] presents a particle filter with optimal discrete density for hybrid state estimation. Particle filters are combined with hidden Markov model (HMM) filter. The objective is to reduce the dimension of the state spaces which are approximated by Monte Carlo sampling and the number of particles. The combination can be an efficient method for hybrid state space models where the dimension of the discrete variables is relatively low.

3.4.3.2. Distributed State Estimation Approaches

Although there are suitable solutions available for specific types of state estimation problems, it is still a challenge to efficiently and effectively implement them in distributed environments. Additional concepts and mechanisms are required to successfully exploit and integrate the individual capabilities of multiple nodes in such a network of system based on a distribution state estimation approach.

Rosencrantz *et al.* [RGT03] present an architecture for decentralized state estimation addressing a distributed surveillance scenario consists. They describe a strictly decentralized approach based on particle filters in which only nearby platforms exchange information and an interactive communication protocol selects data for exchange with the overarching goal to maximize information flow.

Hu and Evans [HE04] introduce a sequential Monte Carlo Localization method for a setup with a small number of seed nodes that know their location and other nodes that estimate their location based on location messages that they receive. One of their results is that mobility can improve accuracy and reduce costs of localization. Huang *et al.* [HZ07] study how different types of sensor data can be incorporated into a common localization framework. They claim that they provide the first particle-filtering framework that allows heterogeneous sensor types to solve localization problems.

An approach for collaborative sensor fusion based on mobile agents is presented by Biswas *et al.* [BQX08]. In their model mobile agents migrate from node to node following a specific itinerary fusing information/data locally at each node. Following this strategy intelligence is distributed throughout the network edge and overall communication cost is reduced. The performance of the mobile agent-based approach is evaluated and compared to a traditional client/server-based computing model.

Chang [Cha08] evaluates the performance of different distributed fusion algorithms. He focuses on scalable fusion algorithms and conducts an analytical performance evaluation to compare their performance under different operating conditions. In particular, the performance of channel filter fusion, Chernoff fusion, Shannon Fusion, and Battachayya fusion algorithms are evaluated.

The consistency problem in decentralized data fusion based on particle filters is still an open challenge. In a distributed environment particles from sample set of two different particle filters (e. g., local and remote) do not have the same support. Therefore one particle set has to be transformed into a continuous distribution. The samples of the second particle filter can be drawn based on this distribution in order to obtain new importance weights [OUR*05].

Ong *et al.* [OUR*05, OUR*06] use two different methods for the transformation of sample statistics maintaining accurate summary of the particles: a Gaussian Mixture Model (GMM) and an approximation of particles by Parzen representation. Common information is removed in the fusion process in order to avoid over-confident estimates due to "double counting". They obtain better fusion results for the Parzen representations than for the GMM. In a subsequent paper Ong *et al.* [OBDWU08] propose a decentralized particle filter based approach for multiple target tracking. They address the problem of common past information in discrete particle sets by a new approach that removes common past information in a division operation. The proposed approach is limited to tree connected networks.

Another approach for distributed state estimation based on particle filters is followed by Kan *et al.* [KPCD12]. They propose an average consensus algorithm for particle based target tracking. They assume that in network of sensor only a portion of sensors nodes actually observe the target. They propose an average consensus algorithm where each sensor tracks the global mean of all local location estimates about the target. They propose a communication model based on gossiping with direct neighbor nodes only.

3.4.3.3. Out-of-Sequence Measurements

In distributed environments measurements coming from multiple sensors typically arrive at the fusion node with some delays. The processing of *out-of- sequence measurements (OOSM)* is a particular challenge in distributed state estimation which is discussed in more detail in the following.

Rosencrantz *et al.* [RGT03] address the integration of OOSM based on re-simulation in their architecture for decentralized state estimation. Re-simulation is performed in order to reduce the variance of importance weights if measurements arrive with some delay referring to an earlier point in time. In this case part of the history of an agent's particle filter is erased and then re-simulated to the current time, incorporating all collected observations at appropriate time intervals.

Mookerjee and Reifler [MR04] address the OOSM problem in a multi-sensor fusion scenario using an optimal reduced state estimator. They provide a simulation example with two sensors: one delivers highly accurate measurements with some delay and the other one provides less accurate measurements but without any delay. Their results show uniform performance improvement in comparison to two traditional approaches.

Many approaches proposed in the literature focus on advanced algorithms to decrease system latency in comparison to measurement buffering approaches. Mauthner *et al.* [MEKB06] provide a comparison of these two concepts by evaluating their effect on the covariance matrices of the state vector. They point out that the OOSM problem may not only be due to varying transmission delays. OOSMs may also be present due to other reasons when deterministic communication protocols are used, for example, with time-triggered measurement mechanisms when multiple sensors have different measurement cycle times. Their analysis shows that the choice of a suitable concept (buffering and advanced algorithms) should be based on the underlying model. In the case that the measurement cycle time of one sensor is a multiple of the cycle time of the other sensor both concept are competitive.

Orguner and Gustafsson [OG08] present a storage efficient particle filters approach for the OOSM problem using auxiliary fixed point smoothers. The proposed solution enables an effective combination of OOSMs with arbitrary delay while minimizing storage requirements. They compare their solution with previously proposed solution in a target tracking scenario based on simulations. Accordingt to their evaluation results the proposed alternatives to existing solutions are promising and reduce storage requirements.

Zhang and Bar-Shalom [ZBS11] address the problem of *multiple OOSMs*. As there are already several algorithms for target state update (optimally or suboptimally) coping with OOSM, there is a need for solutions for multiple OOSMs arriving in arbitrary orders. The straightforward approach for multi-OOSM is the sequential application of an OOSM algorithm. They show, however, that an optimal update for multi-OOSM cannot be guaranteed in this case.

They provide mathematical definitions of out-of-sequence measurement (OOSM) as well as *in-sequence measurement (ISM)* (see examples in Figure 3.11):

Definition 3.6. *Given a measurement $z(k_i)$, if there exists another measurement $z(k_j)$ with $t_{k_j^a} < t_{k_i^a}$ and $t_{k_j} > t_{k_i}$, then $z(k_i)$ is an of out-of-sequence measurement (OOSM). Otherwise, $z(k_i)$ is an in-sequence measurement (ISM).*

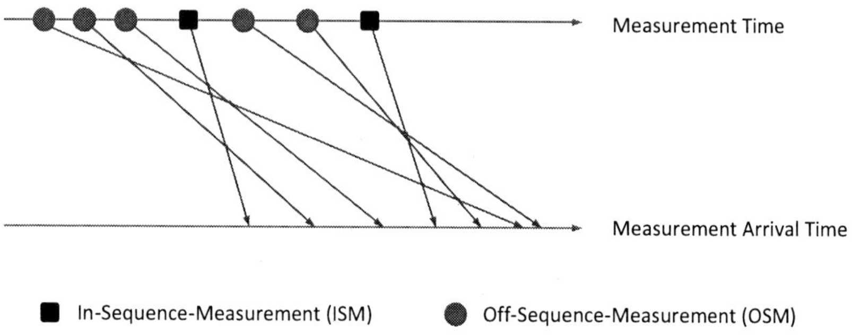

Figure 3.11.: Example Scenario of Out-of-Sequence Measuremens (OOSM) and In-Sequence-Measurements (ISM) (based on [ZBS11])

They also provide formal definitions of *most recent time (MRT)* and *starting time (ST)*. They are based on a newly arriving measurement at a fusion node and the set of measurements that have already arrived before. MRT refers to the latest measurement time out of the set of all measurements that have already arrived at a specific fusion node. ST is the time of the latest measurement (out of the set of measurements that have already arrived) that is older than the newly arriving measurement. These two definitions are the basis for their thorough discussion of single-OOSM – in-sequence OOSM (IS-OOSM) and out-of-sequence OOSM (OS-OOSM) – as well as different types of multi-OOSM (type 1, type 2 and general). The general multi-OOSM scenario is the most challenging where multiple OOSMs may arrive in an arbitrary order, e.g., interleaving with other ISMs.

They propose an optimal solution for the general multi-OOSM scenario based on complete in-sequence information (CISI). This approach also updates all states between OOSM time and most recent time. They present numerical examples showing the optimality of their CISI based methods in various multi-OOSM scenarios.

In a subsequent paper Zhang and Bar-Shalom [ZBS12] present an exact Bayesian solution for OOSM processing with particle filters (called A-PF). It is shown that A-PF is the only algorithm achieving optimal performance obtained from in-sequence processing. However, A-PF has a very high complexity and is therefore only suitable for simple problems or non-real-time applications.

Berntorp *et al.* [BAR12] propose a storage efficient particle filter approach for addressing multiple OOSMs. This solution is based on [OG08] but updates estimates and covariances for all times between measurement time and most recent time when OOSM arrives not only the last estimate. It is an extension of the CISI based approach of Zhang and Bar-Shalom [ZBS11] (described above) to nonlinear systems using it for particle filters. Simulation experiments show that the proposed solution outperforms state-of-the-art particle filter algorithms in some respects while being more storage efficient.

3.4.3.4. Incorporation of Additional Data Sources

There are several object tracking approaches proposed in the literature that also incorporate additional data sources. Moving target indication (MTI) systems that try to discriminate a target against clutter based on radar measurements. Ground MTI (GMTI) approaches are of particular interest within this context as they may incorporate topographical information about the environment of the tracked objects.

Arulampalam *et al.* [AGOR02] present a variable structure multiple model particle filter (VS-MMPF) for GMTI tracking. They propose to use additional information, e. g., road maps and speed constraints, to enhance performance. Roads are modeled as segments represented by two way-points. The proposed concept shows better performance than previously proposed variable structure interacting multiple model (VS-IMM) algorithms. An improved VS-IMM for GMTI tracking is presented by Pannetier *et al.* [PDP08]. There concept includes a stop-move target maneuvering model and additional contextual information. Caicai and Wei [CW09] propose another variable structure interacting multiple model particle filter (VS-IMM-PF) algorithm using road information.

Agate and Sullivan [AS03] present a particle filter based approach for target tracking and identification including road constraints. They chose particle filters as they are not restricted to Gaussian PDFs and can approximate a mixture of continuous and discrete random variables, e. g., target kinematics and target ID. They model the road network as a set of line segments and a set of nodes representing intersections. They evaluate their approach and show the general feasibility of jointly tracking and identifying targets using a particle filters.

Ristic *et al.* [RAG04, chapter 10 "Terrain-Aided Tracking"] outline the concept of Terrain-Aided Tracking. They propose to exploit topographical knowledge of the environment or limitations in the dynamic motion of the target to produce more accurate location estimation results. They present a variable structure multiple-model particle filter that includes terrain conditions by adaptively selecting a subset of modes that are active at a particular time. Their evaluation shows that the target tracking performance can be significantly improved in this way.

Ulmke and Koch [UK06] also propose to exploit additional background information, e. g., road maps and terrain information, to enhance of tracking performance. They present a Bayesian approach to consistently incorporate this kind of information, including winding roads and road networks. They list the following types of a priori information that could be used to enhance GMTI performance: a realistic model of GMTI sensor and measurement process, realistic model of target dynamics, road-maps and topography information as well as tactical background information.

Mertens and Ulmke [MU08b, MU08a] present in a subsequent paper an advanced GMTI approach incorporating additional context information and a refined sensor model. They incorporate a more realistic sensor modeling including clutter notch of the sensor. This way the tracking performance is further improved in respect to track precision and track continuity.

Streller [Str08] presents a road map assisted GMTI algorithm supporting junctions as well as curves. A Gaussian sum algorithm is used within a variable structure multiple model approach using Kalman filtering for system update.

Opitz *et al.* [ODK11] present nonlinear tracking techniques based on GMTI and MTI/MTD radar data. They additionally use relevant constraints based on a topography model of the environment. The road map model is given by Vector Map (VMAP) data that was converted into the ESRI Shapefile format. The algorithms were implemented and could be used in real time environment systems. The implemented algorithms were evaluated and the evaluation results are promising. A quick and dynamic map storage solution is still to be developed in order to be able to handle the huge amount of roadmap data.

Part II.

Research on Robust Situation Awareness

4. Cross Data Analysis

Robust situation assessment and awareness require timely and accurate information on the disposition of the network and the potential and putative threats such as nodes suspected of being compromised. Wrong data, e. g., injected by malicious nodes, should be detected and the systems should be robust against erroneous data, e. g., due to sensor failures. Besides the analysis of available sensor data this may also imply the incorporation of information on the mission to be accomplished and of additional knowledge sources. As node mobility and related aspects pose particular challenges in MANET environments the main focus within this dissertation in on location-related aspects.

The objective is to enable the exploitation of all location-related information available to provide comprehensive awareness of the current situation in a potentially hostile environment. The broad set of available data sources in tactical MANETs is explored and a general model for exploiting cross-relationships for consistency checks is developed. As a proof of concept location-related data sources are analyzed and modeled and cross-relationships are exploited for comprehensive and effective detection of malicious behavior.

The proposed cross data analysis concept exploits cross-relationships of different sensor data and knowledge sources that refer to a specific target of assessment, e. g., network participants or other resources (see figure 4.1). A broad set of partially overlapping data sources is available in tactical MANETs including

- direct sensor data, e. g., GPS coordinates,
- indirect sensor data, e. g., radio signal strength for distance estimation,
- mission-specific knowledge sources, e. g., mission profile, mobility patterns, and
- general knowledge sources, e. g., topography model of the environment.

Cross-relationships between these data sources are investigated, modeled and exploited for *consistency checks* of related data sources to detect false or malicious data.

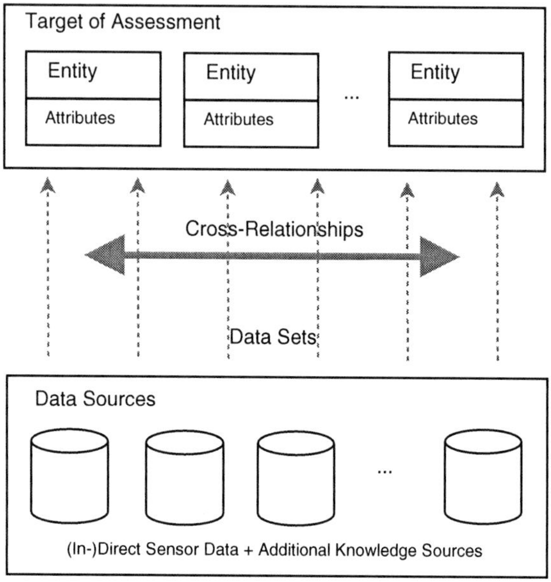

Figure 4.1.: Generic Model of Cross Data Analysis: Different Data Sources Providing Various Data Sets, Contained Cross-Relationships and Targeted Entities of the Assessment

Cross relationship should of course not only be considered between data that is collected on one node but also in respect to data provided by other nodes. The trustworthiness of remote data sources is addressed in chapter 5 based on a cooperative trust assessment approach.

The developed cross data analysis concept is applied to a specific type of situation awareness information: location related data sources. The available data source within this context and their cross-relationships are analyzed, modeled and exploited for the detection of malicious behavior. Within this context the resource limitations of tactical MANETs have to be considered. Efficient models and detection mechanisms are required that can be executed in real time on resource-constrained mobile devices.

In the following we give an overview of data source generally available in tactical MANETs, we discuss potential overlaps of data sources and we analyze different kinds of cross-relationships which may exist even in the case of non-overlapping data sets. Based on this discussion we show how consistency checks can be exploited in cross data analysis to increase robustness of situation awareness and apply the proposed mechanisms to location related data sources as a proof of concept.

4.1. Data Sources

In this section we examine the data sources that are available in tactical MANETs. They are presented starting with the most specific data sources that are directly related to the operational team which are most valuable to for the situation assessment (direct and indirect sensor data). We continue with the mission objectives and finally present general knowledge source which may be utilized in order to improve the data analysis:

- *Direct sensor data:* Direct measurements that are directly used for the analysis, e. g., GPS coordinates.
- *Indirect sensor data:* Data that can be derived from direct sensor measurements, e. g., distance estimation based on radio signal strength measurements.
- *Mission-specific knowledge sources:* Information that is related to the specific mission carried out by the tactical team.
- *General knowledge sources:* General information sources, e. g., maps or weather forecasts, that may be incorporated into the analysis process.

The different types of data sources are presented in more detail in the following sections.

4.1.1. Direct Sensor Data

The primary purpose of a sensor is to provide measurements, e. g., GPS coordinates, that can directly be used for the analysis within a specific application scenario, e. g., location tracking. Figure 4.2 shows a generic model of a sensor providing a digital representation of real world properties to an application based on (physical) measurements.

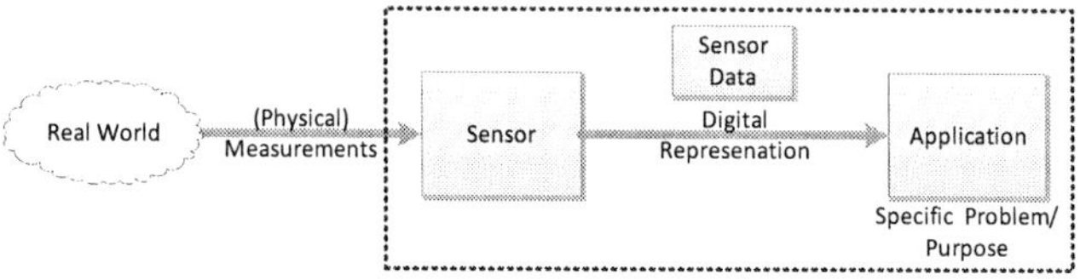

Figure 4.2.: Generic Sensor Model

Sensors are typically developed within a needs-oriented process. They are designed and implemented in order to solve a specific application problem. Therefore the most suitable sensing method has to be determined for a specific application scenario.

In the context of tactical MANETs local sensor data comprise, besides the information that is available on the different network levels of the communication system (e. g., routing information), also additional data provided by specifically tailored external sensor. Additional sensor data can either be retrieved by exploiting characteristics of the communication system that are not necessarily required and evaluated for routing and packet forwarding (e. g., radio signal strength or WiFi chipset fingerprinting) or by integrating external sensors for specific purposes into these systems [AE08a, AE08b]. External sensors may include geolocation sensors (such as GPS), marker-based systems based on known reference points (e. g., at a vehicle), mechanisms to estimate node distances (e. g., distance measurement based on signal propagation times) or human input.

Communication protocols are typically modeled as finite-state machines that represent the current status of the protocol. Within this context we also consider state information, including e. g., connections status and routing tables, as sensor data.

Our main focus is on sensor data that is specific to MANETs and the challenges in tactical environments. There is a lot of previous work available about sensor data in classical wired and wireless networks, e. g., for intrusion detection systems. Therefore we do not discuss in further detail host based data sources such as audit trails and system logs.

4.1.1.1. Sensor Data on Different Communication Layers

In the following the sensor data that is available in a tactical MANET is described based on the four abstraction layers of the TCP/IP model (see RFC1122 [Bra89]): Link Layer, Internet Layer (internetworking), Transport Layer (host-to-host), and Application Layer (process-to-process). This model is used instead of the seven-layer OSI Reference Model [ISO94] as the implementation of a tactical MANET is typically based on common COTS hardware and the Internet Protocol suite.

Link Layer Typically radio interface on 802.11 (WLAN), maybe additional protocols such as ARP (Address Resolution Protocol) and RARP (Reverse Address Resolution Protocol)

- Radio transmission/propagation characteristics (see section 3.2.4.2 for an overview of radio propagation models)
  - transmission power
  - propagation time
- WLAN card/WiFi chipset fingerprinting
- neighborhood watch/"over-hearing" of transmission of other nodes to check if they forward packets as expected
- MAC information (OSI level 2)

Internet Layer Typically IP (Internet Protocol) plus a MANET routing protocol (e. g., OLSR or AODV) and additional management protocols such as ICMP (Internet Control Message Protocol)

- IP packets (including e. g., packet origin, destination, payload)
- routing packets
- local routing information, routing tables

Transport Layer Typically UDP (User Datagram Protocol), potentially also TCP (Transmission Control Protocol)

- UDP (and TCP) packets (including e. g., packet origin, destination, payload)

- for TCP: (open) connections

Application Layer Typically Web based information systems, audio communication, chat and proprietary application protocols

- Web based information systems using HTTP (Hypertext Transfer Protocol) or HTTPS (Hypertext Transfer Protocol Secure)
- audio communication and chat
- proprietary application protocols, e. g.
  - disaster management system, e. g., providing operational picture, description of disaster, disaster area, (location of) victims
- file transfer and mail exchange using FTP (File Transfer Protocol) and SMTP (Simple Mail Transfer Protocol)
- management protocols SNMP (Simple Network Management Protocol), DNS (Domain Name Service)

4.1.1.2. Additional Sensors

Besides the sensors that are available on the different communication layers there are additional sensors available in tactical MANETs.

- geolocation sensor, typically a GPS (Global Positioning System) sensor
- accelerometer
- battery charge indicator
- environmental conditions, e. g., sensor for temperature, brightness, humidity or radiation

4.1.2. Indirect Sensor Data

Data provided by a sensor may deliver indirect hints related to other data sources besides their actual purpose. Direct sensor data can be further processed and analyzed in order to derive indirect sensor data. This data can be used for secondary applications and exploited for cross data analysis as shown in figure 4.3.

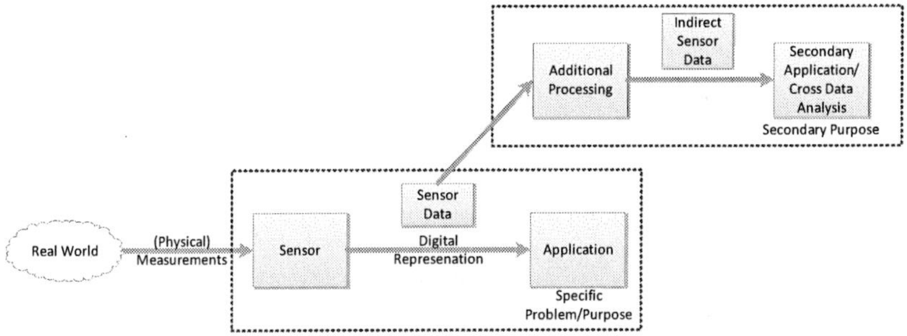

Figure 4.3.: Extraction and Utilization of Indirect Sensor Data

In the following we give three examples for this kind of indirect sensor data.

Distance estimation based on radio signal strength The accuracy of distance estimation based on radio signal strength is affected by various effects such as reflection, diffraction, absorption and interference caused by multipath radio propagation. The reliability and accuracy of measurements therefore strongly depends on the environmental conditions. The relationship of the radio signal strength to other parameters, e. g., node distances, can be described using different radio propagation models such as free-space or two-ray ground model [Rap02].

Wireless Network Card Fingerprinting Individual wireless devices (chipset as well as wireless driver) can be detected by wireless network card fingerprinting mechanisms. These fingerprinting algorithms exploit small but detectable differences in the individual implementation and instantiation of the wireless device, e. g., clock skews or implementation differences of active scanning algorithms. Precise time measurements and statistics allow a timing analysis that is able to distinguish between individual wireless chipsets. These mechanisms do not require any specialized hardware and can be implemented without any modifications to and cooperation of the fingerprinted devices.

Activity Classification based on Energy Consumption Another example is a sensor for the battery status of a mobile device. This information can be used to derive a profile for the energy consumption over time. This information directly reflects the data processing and/or communication intensity. It may be used to classify the activity status of a mobile device and to determine in which activities it is involved.

The dynamics of the system and the environment have to be considered for indirect sensor data as well. Re-calibration and updates may be required for the additional processing steps from time to time.

4.1.3. Mission-Specific Knowledge Sources

There are mission-specific knowledge sources that can be accessed for cross data analysis in tactical environments. This includes a mission profile for the current operation that is in action:

- mission type and objectives
- operation plan/procedures
- operation area
- time schedule
- team composition
- team members
- available equipment

Besides that there is background knowledge for specific mission types or categories:

- standard operation plans and procedures, typical tactical mobility patterns
- standard teams
- general model of known trust relationship
- categories and types of used equipment (e. g., hardware devices, transmission power, battery power)

Some aspects may also dynamically change, e. g., the current team composition. This includes planned actions, e. g., based on mission plans and procedures, but also spontaneous changes due to other reasons. The network may be partitioned, a group of nodes may get out of reach of the other nodes and nodes may (re-)join the network.

4.1.4. General Knowledge Sources

In addition to tactical knowledge sources generally available knowledge sources may be used for cross data analysis:

- topography model of the environment: geospatial data and maps
- weather forecast
- time of sunrise and sunset; moonrise, moonset and moonphase
- other (political) news about the current situation (e. g., uprisings, protests, terrorist attacks)

4.2. Cross-Relationships and Consistency Checks

Important aspects for cross data analysis are partial overlaps of available data sets and cross-relationships between data sets. Additionally sensors may deliver indirect hints related to other data sources besides their actual purpose that can be exploited for cross data analysis.

4.2.1. Overlaps of Data Sources

Available data sources can be categorized according to the following three properties to get a clearer few on potential overlaps:

- *Time:* This is the moment in time or time span that a specific data element (e. g., a sensor measurement) refers to.
- *Space:* This is the location or area that a specific data element refers to, e. g., location of measurement, observation area.
- *Subject:* This is the subject or topic that a specific data element refers to, e. g., an object, a victim or a team member.

Figure 4.4 shows an abstract visualization of this categorization as a three dimensional space. A real model of data source model should of course be based on a multidimensional space where space and subject need to be further elaborated.

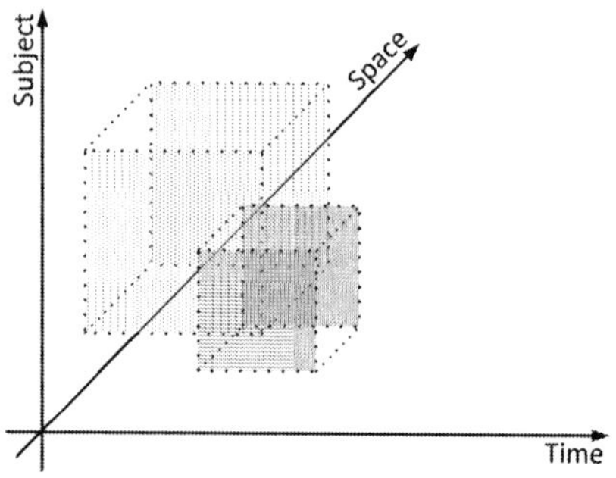

Figure 4.4.: Overlaps of Data Sources in Respect to Time, Space and Subject

Overlaps in data sets may not be obvious due to various reasons. The same (or related) data may be stored in different ways by different applications or mobile systems. Therefore, data alignment is required in several respects in order to detect overlaps and compare the related data sets. This includes, for example, the following aspects:

- *Data Storage and Encoding:* Data has to be stored and accessed in the memory of the computing devices on a lower level, including data alignment as well as data structure padding. Mobile devices and applications may use multiple encoding schemes, e. g., ASN.1 or XML.
- *Data Structure and Formatting:* Data set have to be organized in some way in order to represent more complex data structures. This includes, for example, the grouping and ordering of data elements.

Additionally, (simple) transformations may be required in order to be able to directly compare multiple data sets. This includes conversions and mappings, e. g., due to different measurement units, sensor calibrations or varying sensor data semantic.

4.2.2. Cross-Relationships

Cross-Relationships and cross-dependencies may not only be available for direct overlaps of data sources (as presented above) but also between disjoint data sets due to specific cross-relationships. They are typically based on some kind of causal dependency.

Nevertheless temporal and spatial aspects are important in this context. The incorporation of additional knowledge sources, e. g., mission information or a topography model of the environment, may help to model and exploit these cross-relationships.

The following shows an attempt to give a *categorization* based on their nature:

- laws of physics: due to physical effects
- human rules and laws: sequential order of actions and typical procedures in tactical MANETs (e. g., including chain of command)
- human cognition: human (mental) reaction to cognitive activity
- other typical patterns and sequences

Physical cross-relationships might be most important as they have a clear foundation (laws of physics) and are generally applicable to all areas. Some of them may only take effect in the direct neighborhood: (near field) whereas other effects may also apply for long distance effects.

- electromagnetic effects, including radio propagation
- light emission, reflection, refraction, absorption
- classical thermodynamics, e. g., modeling ideal gas
- basic physical laws of classical mechanics: mass inertia, conservation of energy, conservation of momentum
- chemical effects

Additional some *specific, unique characteristics* may be used for identification of individual instances. Two instantiations of a real-world object are never exactly the same; an individual objects always differs from another implementation or instantiation (at least at a very detailed level).

4.2.3. Consistency Checks

In this section we combine the analysis of data sources and their overlapping and the analysis of cross-relationships between data sources (see sections 4.1 and 4.2). Overlaps and cross-relationships between data source can be exploited for consistency checks in order to check the validity and cleanness of the available data.

For example, mission profiles can be used to verify if specific actions are aligned with the mission objectives. Scenario specific mobility models can be exploited to check the plausibility of node localization and movements. Geospatial data and maps can be taken into account and correlated with other data, e. g., using an advanced radio signal propagation model based on a topographic model of the environment.

4.2.3.1. Plausibility Checks

Plausibility checks are done in different application areas, e. g., for credit approval. A bank takes the data that is provided by a credit applicant and applies plausibility checks in order to analyze the credibility of the provided data. Additionally data is collected by other means in order to do also some background checks.

Applications of plausibility checks in information technology are database management and data warehousing. When new data is imported and integrated into a database system the plausibility of this data is verified in order to ensure a high quality of the data.

This principle can also be applied to cross data analysis in tactical MANETs for a rating of the available information. The basis for the plausibility checks is existing background knowledge about the system and the environment and a modeling of relevant constraints (physical, etc.). The available data sets are analyzed to determine their plausibility and to detect obviously incorrect values.

There are simple *plausibility checks* that can be directly applied to available data set, e. g., checking of data type and encoding checks or range values and limits. The checking of constraints may reveal malicious or erroneous data in a simple way. More sophisticated *consistency checks* may cross check several data records. This may involve the analysis of data provided by multiple entities, data sources or systems.

The focus of cross data analysis is on the latter type of checks which involves more than one data set. These checks will be referred to as "consistency checks", whereas "plausibility checks" are only used for simple check of the internal sanity of a data set in the following.

Consistency checks are discussed in more detail in the following section.

4.2.3.2. Inconsistency and Consistency

Based on the definitions given in [Stu99, Ngu08] (cf. section 3.2.2) we define *inconsistency* and *consistency* of data in the following way:

Definition 4.1. *An* inconsistency *exists if two or more available data sets (provided by sensors and/or knowledge sources) yield at least one contradiction under the given interpretation. The concrete data which yield a contradiction is called* inconsistent data.

Definition 4.2. Consistency *of data is the absence of contradictions, i. e., all available data is compliant and does not yield any contradiction under the given interpretation.*

These definitions are based on the term contradiction, which is defined in the following:

Definition 4.3. *A* contradiction C *consists of the following tuple* (s, D, r)*:*

- *The* subject of contradiction s *is the part of the real world that the inconsistency refers to. It is part of the target of assessment and may consists of one or more entities with specific attributes.*
- *The* conflicting data set D *is a set of data elements* $d_i \in D$ *which directly or indirectly refer to the subject of contradiction and may belong to different data sources.*
- *The* cross-relationship r *defines a relationship between the data sources that the conflicting data sets* $d_i \in D$ *belong to, e. g., represented by formulae or (existence of) relationships.*

The criterion for contradiction is that the cross-relationship r *that does not hold between the data elements* $d_i \in D$.

The target of assessment may be, for example, some team members (represented by network nodes), some resources or other people involved (e. g., victims, enemies). Conflicting data sets could be node locations and available wireless connections and the cross-relationships may be given by an appropriate radio propagation model.

A *consistency check* is used to analyze data sets from different data sources in order to find contradictions. Cross-relationships are exploited in order to find conflicting data sets.

4.2.3.3. Consistency Check Example

Classical methods to detect active attacks on MANETS are typically based on the exploitation of routing and packets. In the following we give an example how the concept of cross data analysis can be applied and how cross-relationships of geo-positions and routing information can be utilized for the detection of wormhole attacks in tactical MANETs.

In order to successfully produce and exploit a wormhole, two collaborating malicious nodes X and X' establish a tunnel, outside the normal network, creating a direct connection between them (see figure 4.5). This way they achieve a fast and good connection that attracts a lot of network traffic which adds significant control and power to the malicious nodes (cf. [EB06, BETJ06]). This shortcut is named after a wormhole as it mimics this hypothetical physical phenomenon.

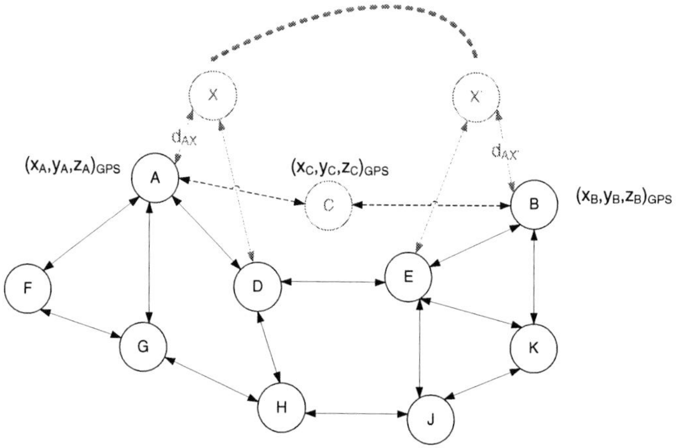

Figure 4.5.: Wormhole Attack by Attacker X and X' Acting as Legitimate Network Node C

A wormhole itself does not have to be negative for the network [EP09]; the potential to maliciously use this channel makes it dangerous. Attackers use wormholes in MANETs to make their nodes appear more attractive (with perceived faster transfer times) such that more data is routed through their nodes, e. g., for eavesdropping. Additionally, they may cut the wormhole connection at a specific critical moment during an operation in order to get a tactical advantage due to temporary network disruption.

The attackers must first get to know to the position of the nodes to be attacked. They act as legitimate network node, for example, using the identity and credentials of a compromised node C. Two cooperating attacker nodes may be located at two different locations (X and X'), which, however, should remain hidden from other network nodes. They appear to the attacked nodes A and B as a single network node with plausible geo coordinates (x_C, y_C, z_C) and may therefore pretend to be located somewhere between them.

Such an attack can hardly be detected using conventional attack recognition methods as each type of data sets provides a coherent view of the network when analyzed on its own. Cross data analysis offers an alternative possibility for attack detection in this case. Cross-relationships of geo-positions and measured distances between nodes can be exploited for the discovery of contradictions between different types of data sets enabling attack detection.

The following cross-relationship must hold between geo-positions of network nodes and measured distances, which may be estimated based on signal propagation time or signal strength:

$$d_{AC} \stackrel{?}{=} d_{AX} = |AX| = \sqrt{(x_X - x_A)^2 + (y_X - y_A)^2 + (z_X - z_A)^2)}$$

$$d_{BC} \stackrel{?}{=} d_{BX'} = |BX'| = \sqrt{(x_{X'} - x_B)^2 + (y_{X'} - y_B)^2 + (z_{X'} - z_B)^2)}$$

Additionally the corresponding cross-relationship can be analyzed in a cooperative approach by other neighbor nodes (e. g., D and E) as well:

$$d_{DC} = |DC| \stackrel{?}{=} |DX| = \sqrt{(x_X - x_D)^2 + (y_X - y_D)^2 + (z_X - z_D)^2)}$$

$$d_{EC} = |EC| \stackrel{?}{=} |EX'| = \sqrt{(x_{X'} - x_E)^2 + (y_{X'} - y_E)^2 + (z_{X'} - z_E)^2)}$$

All of these cross-relationships between geo-position measurements and estimated distances have to be consistent, otherwise there is a clear indication for malicious actions or node failure.

This is an example how cross data analysis may unmask even a sophisticated attacker. Cross-relationships between multiple data sources can be used for consistency checks revealing contradictions even if attackers disguise their malicious behavior.

4.3. Cross Data Analysis of Location Related Data Sources

In this section we apply the concept of cross data analysis described above to location related data sources as a proof of concept. We incorporate direct location sensor data (GPS sensors) and sensor data that indirectly reveal information about node locations (radio signal strength and routing data). A topography model is used as additional knowledge source and an enhanced radio propagation modeling incorporating obstacles is developed for modeling the cross-relationships between node locations and radio signal characteristics. Additionally movement patterns and mobility model can be used for estimating and verifying node locations.

The overall objective is to increase the robustness of the situation awareness system based on cross data analysis of all location related data sources. For these purposes, we develop several consistency check algorithms exploiting cross-relationships in order to detect active attacks on tactical MANETs.

4.3.1. Location Related Data Sources

The MANET is modeled as a directed graph $G = (V, E)$, where V is the set of vertices (nodes) and E the set of edges (symmetric or asymmetric links). Individual network nodes are denoted with upper case letters and $N = |V|$ is the known and fixed number of mobile nodes. E is a subset of all ordered pairs created by the Cartesian product $V \times V$ of all vertices.

In the following we define a mathematical model $M = (V, E, D, R, H, S)$ of location related direct and indirect sensor data. Based on a MANET with N nodes, four $N \times N$ matrices D, R, H and S are defined which cover node distances, route distances, next hops and received radio signal strengths, respectively. Then a topography model is presented as an additional knowledge source which will serve as a basis for the topography based radio propagation model which is presented afterwards.

4.3.1.1. Direct Sensor Data: GPS and Routing Tables

The most important location related data source is of course provided by the geolocation sensor (GPS). Additionally, routing tables indirectly provide information about node distances and hence node positioning.

Node Positions and Node Distances Node Positions can easily be determined using a GPS [HWLC04] receiver, typically integrated in tactical mobile devices.

Definition 4.4. *The geographic coordinates of a network node i are denoted as position* $p_i = (x_i, y_i, z_i)$*. The distance matrix D contains the distances between individual network nodes determined using their GPS coordinates, where* $d_{ij} \in \mathbb{R}$, $i, j = 1, \ldots, N$ *is the distance from node i to node j.*

The distance between two network nodes i and j, i.e., an entry d_{ij} of the distance matrix, can be calculated based on the GPS data provided by nodes i and j using the following equation:

$$d_{ij} = \sqrt{(x_i - x_j)^2 + (y_i - y_j)^2 + (z_i - z_j)^2}.$$

The distance matrix D is symmetrical since the distance from node i to node j is equivalent to the distance from node j to node i.

Routing Information In the following two additional matrices are defined to model routing information.

Definition 4.5. *The route distance matrix R contains the number of hops between individual nodes.* $r_{ij} \in \mathbb{N}$, $i, j = 1, \ldots, N$ *is the number of hops of the route from node i to node j.*

The route distance matrix R is compiled using routing information provided by the individual network nodes. The entry r_{ij} refers to the route to node j that is known by node i as determined by the routing protocol. For simplicity, multi-path routing protocols are not considered.

Usually a route requires the same number of hops in both directions, based on symmetrical connections as commonly used in MANETs. Nevertheless it is possible that the route from node i to node j runs through a greater or lesser number of intermediate nodes than the route from j to node i. Particularly for reactive protocols the route distance matrix R may contain asymmetrical entries since a route from node i to node j may be known, but not a reverse route from node i to node j: $r_{ij} \neq r_{ji}$ for $i \neq j$.

Definition 4.6. *The next hop matrix H contains next hops of all known routes by a network node. $h_{ij} \in N$, $i, j = 1, \ldots, N$ is thus the next hop neighbor node of node i regarding the route to node j.*

Using this matrix together with the route distance matrix R, complete routes between two network nodes can be computed. The next hop matrix as well as the route distance matrix can be extracted from routing information of the individual network nodes. This matrix is generally asymmetrical, i. e., $h_{ij} \neq h_{ji}$ for $i \neq j$ but symmetrical entries may be common in specific network scenarios. Symmetrical entries may occur for example if 2-hop neighbor nodes communicate using a common neighbor node.

4.3.1.2. Indirect Sensor Data: Radio Signal Strength

As described in the consistency check example (see section 4.2.3.3) the attacker need to disguise their true geo coordinates and/or falsify them to successfully implement a wormhole attack. This can be countered if alternative methods for position and distance estimation of mobile nodes based on indirect sensor data are available. The geo coordinates (x_A, y_A, z_A) and (x_B, y_B, z_B) of two network nodes A and B, which have been identified e. g., via GPS, are directly correlated with their distance d_{AB}, which may be determined in an alternative way using radio propagation characteristics (cf. figure 4.6).

Alternative ways of distances estimation using indirect sensor data, e. g., based on radio signal strength or signal propagation time, can help to verify the correctness of transmitted GPS data and to identify manipulation attempts. Furthermore unintentionally false positioning data caused by misconfiguration or hardware failure can be de-

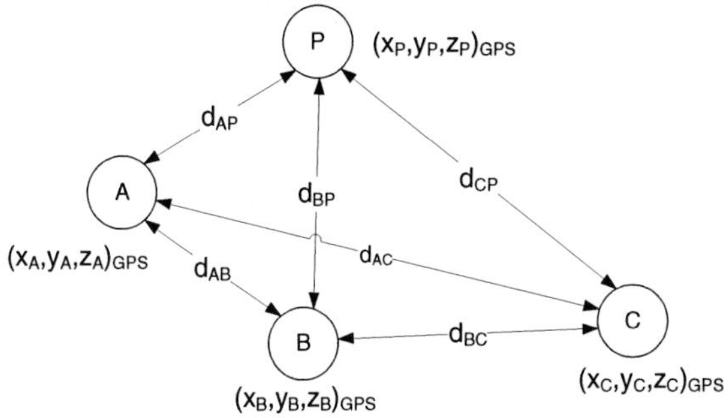

Figure 4.6.: Absolute Geo Positions and Distances Between MANET Nodes

tected. Additionally, alternate positioning methods may provide more accurate data than GPS in some environmental setups.

Radio Signal Strength Radio signal strength values may be used for distance estimation between to network node.

Definition 4.7. *The radio signal matrix S contains radio signal strength measured by each individual node, where $s_{ij} \in \mathbb{R}$, $i, j = 1, \ldots, N$ is the strength of the radio signal of node j received by node i.*

Matrix values s_{ij} reflect as well the relative distance of two nodes, which (if not obstructed by thick walls or other strong shielding obstacles) will show strong radio connection to each other if close together and no radio signal at all if far apart. The maximum radio range is typically a few hundred meters in MANETs. Radio signal strength constantly changes due to node movement and is does not have to be symmetrical between two communicating nodes, i. e., $s_{ij} \neq s_{ji}$ for $i \neq j$.

The relationship of the radio signal strength and node distances, can be described using different radio propagation models such as free-space or two-ray ground model [Rap02]. However, these models are very simplistic and more elaborate propagation models would be required in order to incorporate also reflection and scattering effects. Signal strength also depends on the transmission power of the sending nodes. For simplicity, in our cooperative tactical setup all nodes are expected to have equal transmission power. If a network node (temporarily) decreases its transmission power, e. g., to save energy, this information will also be shared with the other network nodes.

In order to apply positioning consistency checks using alternative distance estimation methods based on radio signal characteristics an advanced radio propagation model is needed. To model the terrain and thus the propagation of radio waves several types of propagation effects can be specifically considered including but not limited to: obstacles between nodes, reflections on ground or solid objects and scattering effects. Therefore we develop a radio propagation model (cf. section 4.3.2 on the next page) based on a topography model of the environment to enable the exploitation of these cross-relationships.

4.3.1.3. Additional Knowledge Source: Topography Model

A topographical model is used as an additional knowledge source which provides the basis for the advanced radio propagation model. The topography model enables the detection of obstacles between nodes, the consideration of reflection effects on solid objects, and diffraction effects.

The ground in our proposed topography model is assumed to be almost flat. To enable the detection of obstacles and to calculate reflection and diffraction factors the topographical model must describe the location of buildings and hold additional information about the objects' surface structure. The surface structure of topographic objects is assumed to be almost constant and the objects need to be large in relation to the transmission frequency (i. e., objects' components need to be substantially bigger than 10 cm). Taking this assumption into consideration all topographic objects are described as polyhedrons, where the smallest components of an object are the faces of the polyhedron. Every face stores one permittivity factor that is used for reflection and diffraction calculations.

Standard shape files [Esr98] used in geospatial models are employed to represent topographical data. As shape files store objects as polygons with an additional parameter for height, the corresponding polyhedra consist only of vertical faces. This limitation to vertical polyhedron faces represents an additional constraint on the model, restricting it to a 2.5D instead of a 3D model.

4.3.2. Topography Based Radio Propagation Model

Highly accurate models of RF (radio frequency) signal propagation exist but they require considerable computational resources and are hence unsuitable for incorporation into *real-time* protocols, particularly on *resource-constrained* platforms such as MANET nodes. We therefore develop a simplified radio propagation model which takes the position of nodes as well as topographical information into account but does not incorporate a comprehensive model of physical effects.

The *topography based radio propagation model* presented in the following was developed within the Diplom thesis supervised by me of Steffen Reidt [Rei06] and the results were further elaborated and published in a joint paper [REKW11].

To model electromagnetic waves and their interaction with terrain particularly in built-up areas a topographical model is used as basis for the ray-optical propagation model which represents the environment, e. g., buildings, streets. The propagation model does not only incorporate effects of direct line of sight transmission but also reflection and diffraction effects which are highly significant e. g., in urban operations. Scattering effects of vegetation is not taken into consideration as its inclusion in a ray-optical computation would be too costly under the constraints previously described.

The model is not as comprehensive as others that also incorporate physical effects, however, we argue that the approximation provided by our model is *sufficient* and *suitable* for the proposed purposes. We will eventually show the advantages of a model which incorporates both geographical disposition of nodes and geolocation information for overall improved performance without the resource intensities required by incorporating all physical effects.

The propagation of electromagnetic waves can be described by solving the corresponding Maxwell equations. In most cases these solutions must be approximated using numerical models. Based on work by Maurer [MFSW04] [MFW05] and lighting techniques commonly used in computer graphics, a ray-optical propagation model can be used to approximate the behavior of electromagnetic waves. According to [MFW05], this approach is defensible provided three primary conditions:

1. The frequency band used is beyond 1GHz.
2. Interacting terrain features are large in comparison to the RF wavelength.
3. The surface structures of individual terrain features are approximately constant.

The first condition is satisfied for the IEEE 802.11 (a/b/g/h) series of standards (which use bands from 2.4 to 2.5 GHz and 5.15 to 5.85 GHz, respectively). The appropriate wavelength of approximately 10 cm at these frequencies is substantially smaller than the topographic objects such as buildings or trees. Furthermore, the model specified in section 4.3.1.3 on page 134 provides only uniform surfaces and does not include additional modifiers such as surface textures, which therefore also satisfies the third condition.

The calculation of receiving power values in this model can be divided into two steps. The first step contains algorithms that determine all intersection points of the wave with the objects of the scene. The second step calculates the reflection and diffraction factors at every point of interaction. In the final step power levels are computed with the help of those factors combined with path length information. The following sections describe the modeling of these two steps based on the underlying topographical model:

Ray Tracing The first step in calculating propagation power within our model is to find all paths from sending to receiving nodes. This requires consideration of three signal paths:

1. Line of sight (LOS) between sender and receiver
2. Paths containing only reflection points
3. Paths containing diffraction points and both reflection and diffraction points

As the effort required depends on the maximum number of reflections for one path, the maximum reflection and diffraction depths can be limited as required by external constraints. The receiving power at a node can be represented by numerous paths. In order to obtain the total receiving power we must first calculate the power transmitted by each individual path. With all individual receiving power values and considering appropriate paths' directions, the total receiving power is then calculated in a final step.

The basic propagation model described below determines power values for line of sight paths. We then augment this estimation by including reflection and diffraction.

4.3.2.1. Line of Sight

Power values caused by a direct line of sight connection between two nodes are obtained from the Free Space or Two-Ray Ground models [FV11]. The Two-Ray Ground formula is only suitable if the distance between sending and receiving nodes exceeds a threshold distance (cf. section 3.2.4.2 on page 68). We therefore use the Free Space model for distances below this threshold. The Two-Ray Ground model assumes horizontally polarized radio waves, which can change with reflection on 3D objects. As objects contained in the topographical model are constructed solely of vertical faces (cf. section 4.3.1.3), it does not affect horizontal polarization. As our model also assumes near-flat ground environments, all conditions for using the Two-Ray Ground model are fulfilled.

4.3.2.2. Reflection

A radio wave is, for the purpose of electromagnetic calculations, defined by its direction and one horizontally and one vertically polarized value. In the following the direction of the respective incident wave is described by $\vec{e}^{\,i}_k$, and the vectors for polarizations are named $\vec{E}^i_h$ and $\vec{E}^i_v$ with the related eigenvectors $\vec{e}^{\,i}_h$ and $\vec{e}^{\,i}_v$. The vectors describing the reflected wave are named with r as upper index instead of i, respectively. Both incoming and reflected waves are thereby described in their own local coordinate system (see Figure 4.7 [REKW11]).

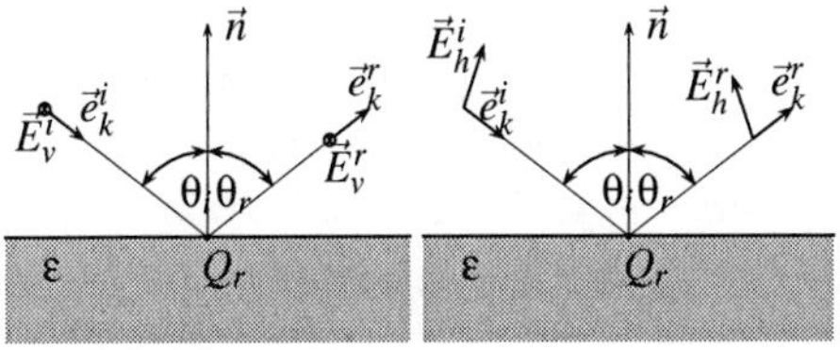

(a) vertical polarisation (b) horizontal polarisation

Figure 4.7.: Reflection on a Flat Surface

ε describes the permittivity of the object, Q_r is the reflection point. Typical permittivity values are: water 81; brick 4.4; sandstone 7.5; glass 4. The angle of incidence and the angle of reflection are described by θ_i and θ_r, with $\theta_i = \theta_r$. For a wave that is reflected on a flat face, the reflection factors R_h and R_v for horizontal and vertical polarization are given by [Rap02]:

$$R_h = \frac{\varepsilon \cos\theta_i - \sqrt{\varepsilon - \sin^2\theta_i}}{\varepsilon \cos\theta_i + \sqrt{\varepsilon - \sin^2\theta_i}} \tag{4.1}$$

$$R_v = \frac{\cos\theta_i - \sqrt{\varepsilon - \sin^2\theta_i}}{\cos\theta_i + \sqrt{\varepsilon - \sin^2\theta_i}} \tag{4.2}$$

Thus the amplitudes $\vec{E}_h^r$ and $\vec{E}_v^r$ located directly at the reflection point Q_r can be calculated as follows:

$$\begin{pmatrix} \vec{E}_h^r(Q_r) \\ \vec{E}_v^r(Q_r) \end{pmatrix} = \begin{bmatrix} R_h & 0 \\ 0 & R_v \end{bmatrix} \begin{pmatrix} \vec{E}_h^i(Q_r) \\ \vec{E}_v^i(Q_r) \end{pmatrix} \tag{4.3}$$

For a node that is located at a reflected ray, the receiving power can now be determined using the Free Space or Two-Ray Ground model, starting at Q_r with $\vec{E}_h^r$ and $\vec{E}_v^r$ as new power values.

4.3.2.3. Diffraction

To calculate the power transmitted by diffraction paths the field can be described in its own coordinate system, which is defined by the wave's direction and the components for vertical and horizontal polarization, similarly to the reflection case. According to the outline shown in Figure 4.8 [REKW11], the incident wave is described by $\vec{e}_k^i$, $\vec{E}_h^i$ and $\vec{E}_v^i$ and the deflected wave by $\vec{e}_k^d$, $\vec{E}_h^d$ and $\vec{E}_v^d$, respectively.

The amplitudes $\vec{E}_h^d$ and $\vec{E}_v^d$ are analogous to equation 4.3 calculated by:

$$\begin{pmatrix} \vec{E}_p^d(Q_r) \\ \vec{E}_s^d(Q_r) \end{pmatrix} = \begin{bmatrix} D_p & 0 \\ 0 & D_s \end{bmatrix} \begin{pmatrix} \vec{E}_p^i(Q_r) \\ \vec{E}_s^i(Q_r) \end{pmatrix} \tag{4.4}$$

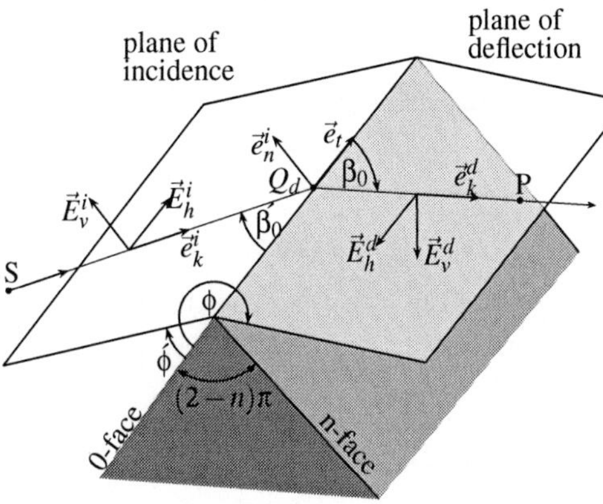

Figure 4.8.: Geometry and Components of the Deflected Field

D_p and D_s are now the deflection factors, which can be calculated as follows [Mau05]:

$$D_p = \frac{e^{-i\frac{\pi}{4}}}{2n\sqrt{2\pi k_0}\sin\acute{\beta}_0}\left(D_0^{\mathrm{ISB}} + D_n^{\mathrm{ISB}} - D_0^{\mathrm{RSB}} - D_n^{\mathrm{RSB}}\right) \tag{4.5}$$

$$D_s = \frac{e^{-i\frac{\pi}{4}}}{2n\sqrt{2\pi k_0}\sin\acute{\beta}_0}\left(D_0^{\mathrm{ISB}} + D_n^{\mathrm{ISB}} + D_0^{\mathrm{RSB}} + D_n^{\mathrm{RSB}}\right) \tag{4.6}$$

Total Receiving Power Knowledge of the reflection and diffraction factors and the path's lengths enables the calculation of the transmission power for every single path. According to [Mau05], the power values are given by

$$P_r = \frac{P_t\lambda^2}{(4\pi)^2}\left|\begin{pmatrix}1\\0\end{pmatrix}^T \mathbf{T}\begin{pmatrix}1\\0\end{pmatrix}\right|^2 \tag{4.7}$$

where **T** is the matrix containing the reflection and diffraction factors and the path's length:

$$\mathbf{T} = \prod_{i=0}^{M}\mathbf{D}_i\mathbf{S}_i$$

M represents the number of reflections and diffractions. The length of the path d is contained in the Matrix $\mathbf{S}_0$:

$$\mathbf{S}_0 = \begin{bmatrix} 1/d & 0 \\ 0 & 1/d \end{bmatrix}$$

$\mathbf{S}_i$ $(1 \leq i \leq M)$ are the matrices of reflection or diffraction factors, which are given by one of the matrices

$$\mathbf{S}_i = \begin{bmatrix} R_p & 0 \\ 0 & R_s \end{bmatrix} \quad or \quad \mathbf{S}_i = \begin{bmatrix} D_p & 0 \\ 0 & D_s \end{bmatrix}$$

from equation 4.3 or 4.4. As the wave is described in its own coordinate system after every reflection or diffraction, the matrices $\mathbf{S}_i$ need to be transformed to the appropriate coordinate system before multiplication. For the purpose of defining the basis transformations $\mathbf{D}_i$, the components describing field strength for horizontal and vertical polarization are named $\vec{e}_{Eh,i}$ and $\vec{e}_{Ev,i}$ for the incident and $\vec{e}_{Ah,i}$ and $\vec{e}_{Av,i}$, respectively. If a path contains for example exactly one reflection, then $\vec{e}_{Ah,0}$ and $\vec{e}_{Av,0}$ describe the polarization directions of the wave at the transmitting node. To compute the reflection factors represented in Figure 4.7 [REKW11], the vectors $\vec{e}_{Eh,1}$ and $\vec{e}_{Ev,1}$ and respectively $\vec{e}_{Ah,1}$ and $\vec{e}_{Av,1}$ are chosen in accordance to the plane of incidence and plane of reflection/diffraction.

Finally, $\vec{e}_{Eh,2}$ and $\vec{e}_{Ev,2}$ describe wave polarization at the receiving node. Consequently $\mathbf{D}_0$ must define the transformation from $\vec{e}_{Ah,0}$ and $\vec{e}_{Av,0}$ to $\vec{e}_{Eh,1}$ and $\vec{e}_{Ev,1}$, and $\mathbf{D}_1$ the transformation from $\vec{e}_{Ah,1}$ and $\vec{e}_{Av,1}$ to $\vec{e}_{Eh,2}$ and $\vec{e}_{Ev,2}$. Matrices $\mathbf{D}_i$ describe a rotation around the wave's propagation direction and thus a transformation in the *plane* defined by $\vec{e}_{Ah,i}$ and $\vec{e}_{Av,i}$, which can be represented by (2,2)-matrices. The vectors $\vec{e}_{Eh,i}$, $\vec{e}_{Ev,i}$ and $\vec{e}_{Ah,i}$, $\vec{e}_{Av,i}$ are already orthogonal in their according coordinate system. Normalizing these vectors allows the definition of the matrices $\mathbf{D}_i$ by scalar products:

$$\mathbf{S}_i = \begin{bmatrix} \vec{e}_{Ah,i} \cdot \vec{e}_{Eh,i+1} & \vec{e}_{Ah,i} \cdot \vec{e}_{Av,i+1} \\ \vec{e}_{Av,i} \cdot \vec{e}_{Eh,i+1} & \vec{e}_{Av,i} \cdot \vec{e}_{Av,i+1} \end{bmatrix}$$

If the path contains no reflections or diffractions, then $\mathbf{D}_0$ is the identity matrix, $\mathbf{T} = \mathbf{S}_0$ and the equation (1) comes out as a special case of equation 4.7. To describe several paths, the matrix $\mathbf{T}$ for j's path is now named $\mathbf{T}_j$. Thus, the total receiving power composed by N paths can be therefore calculated as:

$$P_r = \frac{P_t \lambda^2}{(4\pi)^2} \left| \sum_{j=1}^{N} \begin{pmatrix} 1 \\ 0 \end{pmatrix}^T \mathbf{T}_j \begin{pmatrix} 1 \\ 0 \end{pmatrix} \right|^2 \tag{4.8}$$

4.3.3. Cross-Relationships Between Data Sources

Based on the preliminary work in the sections above we describe in this section the cross-relationships of location related data which will then be exploited to detect active attacks (e. g., black hole or wormhole attack) [EHK*07, ES08, JWK*08].

Node positions directly influence network properties in many ways. Most network properties depend directly and indirectly on node distances and radio connectivity. Therefore, positioning data can be combined with other properties such as radio signal characteristics and routing information to detect active attacks. Consistency conditions can be checked to detect discrepancies compared to normal network operation.

For this purpose additional information sources besides direct positioning data (such as GPS coordinates) are considered to (at least indirectly) draw some conclusions about the distance between individual nodes. Expected communication relationships derived from node localization incorporating radio propagation and topography model are compared to the actual network connectivity graph given by the routing tables. High variances from the typical state of this sensor data during normal network operation may indicate an attack in progress.

4.3.3.1. Location Related Data Sources

In this section attack consistency checks are derived based on the location related data sources described above and their cross-relationships.

Data Sources and Cross-Relationships The following three categories of information are considered for this analysis.

- *Geo coordinates*: These are the globally unique geographic positions of individual network nodes, which can, for example, be estimated using GPS receivers.

From these geo coordinates also geographical distances between specific network nodes can be determined.

- *Routing information*: Each network node maintains its own routing table which contains the next hop and the distance (number of hops) for each destination node. The routing tables of all network nodes may be aggregated and analyzed locally, in a local network cluster or in a central IDS.
- *Radio signal characteristics*: Each node measures and stores the signal strength or received radio communication signals for each direct neighbor node. Additionally it would be possible that nodes exchange additional radio messages to estimate their distances based on signal propagation time (cf. section 4.3.1.2).

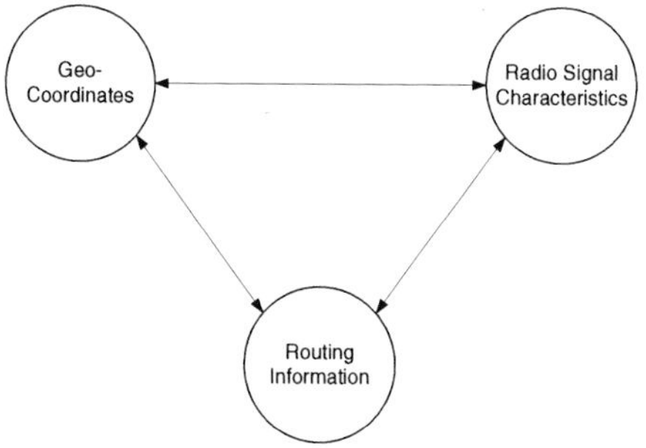

Figure 4.9.: Cross-Dependencies of Geo Coordinates, Routing Information and Radio Signal Strengths

Cross-Relationships Certain cross-dependencies exist between these three categories of information which are shown in figure 4.9. The arrows are supposed to represent the data records that can be (partially) derived from other records. Therefore the following relationships should generally apply if all network notes provide correct information to the IDS and the environmental factors are considered correctly.

- *Geo coordinates and routing information*: Ideally, neighborhood relationships between particular network nodes and therefore potential routes can be determined based on geo coordinates of the network participants and a topographical model of the environment. There may be however several possible alternatives for certain network situations as there might be different valid options. Geo coordinates of network nodes cannot be determined based on routing information. However, an estimate for the upper limit of the distances between certain nodes can be given.

- *Geo coordinates and radio signal characteristics*: Similarly to routing information also radio signal characteristics can (at least theoretically) be determined the geographic coordinates using a suitable radio propagation model. Measured radio signal characteristics, however, do not lead to direct conclusions about geo coordinates of network nodes.
- *Routing information and radio signal characteristics*: Since only direct neighbors can measure their mutual radio signal characteristics, there is no direct correlation between these two information categories (except for direct neighbors, i. e., "one-hop-routes"). The successful measurement of radio signal characteristics of neighboring nodes (that includes his identification resource) should only be the case if a communication link exists and therefore an entry in the routing table. In turn, a communication link to a neighbor node requires a radio communication link and therefore provides radio signal characteristics, in particular, a signal strength that exceeds a minimum noise threshold. Therefore there is at least some minimum threshold for radio signal strength values if nodes are direct neighbors according to the routing information provided.

In order to apply these fundamental, theoretical relationships in practice the identified cross-relationships need to be individually modeled and calibrated. The data sources are influenced by various environmental factors and prone to measurement errors therefore typical parameter values need to be determined for a specific network scenario.

- *Calibration of the radio propagation model*: Suitable parameter values for the radio propagation model can be determined based on measured geolocation (and therefore distance) and corresponding radio signal strength values such that the resulting model approximates the measured values in an optimized way.
- *Estimation of the communication radius*: The routing information show which nodes have a direct communication link and which do not. This can be used to

determine an estimate for the mean and maximum communication radius of network nodes in the current network setup. This may can be done on the one hand by determining the maximum distance between direct neighboring nodes that do have a direct connection in order to determine an upper limit of the communication radius, and on the other hand by also determining the minimum distances between nodes that do not have a direct connection to determine an estimate for a lower limit.

The estimation of these scenario parameters can be performed before the beginning of a particular mission. Additionally, the described parameters could be slowly adapted to changes of the current situation and adjusted during a mission. This includes however the risk that a knowledgeable attacker may trick the system by providing false information in such a way that the system gradually learns wrong system parameters.

4.3.4. Consistency Checks and Attack Detection

In this section we describe how cross-relationships of location related data can be exploited to derive consistency checks. We develop detection algorithms to detect active attacks (e. g., black hole or wormhole attack) based on consistency checks of positioning data, radio signal characteristics and routing information [EHK*07,ES08,JWK*08].

Positioning data, radio signal characteristics and routing information are analyzed and correlated to check their consistency in order to detect anomalies that might indicate attack. For this purpose the following cross-relationships are further investigated:

- *Geo coordinates vs. communication links*: It should be verified whether the calculated distance between two network nodes is consistent with the estimated communication radius.
- *Route lengths (number of hops) vs. geographic distances*: Generally it can also be checked for complete itineraries whether the distance of the communicating nodes is consistent to the maximum communication radius. For this purpose the number of hops of a route is multiplied by the maximum radio communication distance and compared with the calculated distance (based on geo-coordinates).
- *Radio signal strength vs. geo coordinates*: Radio signal strengths calculated based on geo coordinates are compared with signal strength measurements.

Attack Detection Modules based on Consistency Checks In the following sections these cross-relationships are further elaborated to derive three consistency checks and attack recognition methods. The mechanisms were implemented within the Diplom thesis supervised by me of Martin Sommer [Som07] and the results were further elaborated and published in a joint paper [ES08].

The three attack recognition methods are are based on a unified model of the three categories of information (node distances, routing information and radio signal strengths) as defined in section 4.1.1. This matrix based approach allows an effective and efficient development of analysis methods that detect significant changes during a black hole attack which contrast normal network operation.

These matrices provide a very descriptive way to examine node positions and distances, radio signal strengths as well as routing information for detecting possible inconsistencies. An important property of this approach is the consistent modeling of different data sources, i. e., in all matrices at position ij the value of node i in relation to node j is stored.

Another objective of this detection approach is to allow matrices that are not completely filled with values to be analyzed without separate treatment. More information allows a more thorough analysis, but within a dynamic network topology a complete and always up-to-date set of data cannot be expected.

The main focus is to evaluate the feasibility and applicability of attack detection based on node localization and the related data sources in MANETs.

In the following sections three different analysis methods for consistency checks are presented: distance verification, connectivity examination and radio signal strength checking (see table 4.1) [ES08].

4.3.4.1. Cross Data Consistency Check: Distance Verification

This analysis method examines the relationship between node positions and radio signal strength measurements. Generally, nodes that are located closely together can receive radio signals from each other whereas distant nodes have no radio link.

Within a wireless network a maximum radio range can be estimated, referred to as d_{max}. There is a close relationship between the length of a route and the geographical

Table 4.1.: Consistency Checks based on Location Related Data Sources

Consistency Check	Geographical Positions and Distances (D)	Routing Information (R, H)	Radio Signal Strength (S)	Description
Distance Verification	X	X		Verification of consistency of route lengths and geographical distances
Connectivity Examination		X	X	Examination of network topology according to radio connections
Radio Signal Strength Checking	X		X	Radio signal strength estimation based on node geo coordinates

distance of its source and destination node. Based on this information simple inconsistency checks can be accomplished between the distance matrix D and the route distance matrix R.

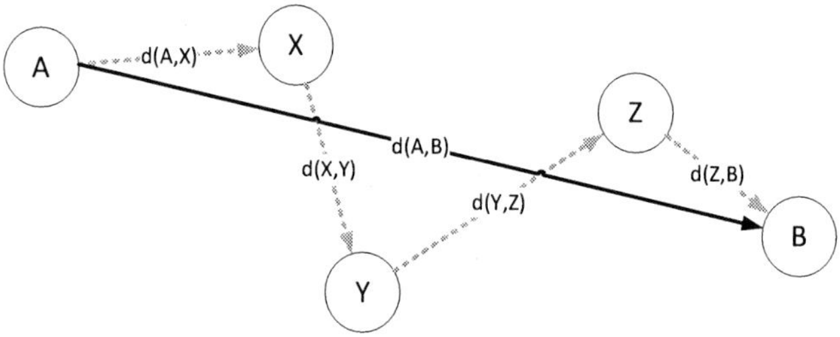

Figure 4.10.: Distance Verification of a Four Hop Route from Node A to Node B

For one-hop entries in the route distance matrix ($r_{ij} = 1$) it is not expected that distance of the related nodes is larger than the estimated maximum radio range ($d_{ij} < d_{max}$).

For example, for direct neighbor nodes A and node B, i. e., $h_{AB} = 1$, it would not be expected that distance of the related nodes is larger than the estimated maximum radio range ($d_{AB} < d_{max}$).

The distance examination is however not limited to the check of direct neighborhood properties. Figure 4.10 shows an example of a distance verification of a route from node A to node B which contains four hops, in this case $d_{AB} \leq 4 \cdot d_{max}$

Generally the following relation can be used for normal MANET operation:

$$\forall i, j : d_{ij} < r_{ij} \cdot d_{max}. \tag{4.9}$$

4.3.4.2. Cross Data Consistency Check: Connectivity Examination

The connectivity examination utilizes the cross-relationship between radio signal characteristics and network topology information. If two network nodes have a direct connection to each other (route length one), they must have a direct wireless connection and the measured signal strength must be above a specific threshold s_{min}. On the other hand, if two nodes have no direct connection, they do not have a radio link.

The following relationship between route distance matrix R and radio signal strength matrix S should be true:

$$\begin{aligned} r_{ij} = 1 \quad &\Rightarrow \quad s_{ij} > s_{min} && \text{for asymmetrical connections and} \\ r_{ij} = 1 \quad &\Rightarrow \quad s_{ij} > s_{min} \wedge s_{ji} > s_{min} && \text{for symmetrical connections, respectively.} \end{aligned} \tag{4.10}$$

Thus it is examined whether a node i, which claims to be a direct neighbor of a node j really possesses a direct radio link. With asymmetrical connections this examination is not effective since the value s_{ij} as well as r_{ij} are delivered by node i and therefore can easily be forged suitably by a potential attacker. However a missing radio signal from node i can indicate some incorrect connectivity information by receiver node j (based on typical symmetrical connection behavior). Equivalently the following condition should hold in cases of symmetrical connections: $r_{ij} = 1 \Rightarrow r_{ji} = 1$.

4.3.4.3. Cross Data Consistency Check: Radio Signal Strength Checking

This analysis method is based on consistency checks between node positioning and routing information. Node geo coordinates of sending and receiving nodes can be used to estimate the strength of the received radio signal. Figure 4.11 shows an example scenario for this consistency check.

Let s_{ij} be the radio signal strength of node j received by node i and $f(p_i, p_j)$ a function to estimate received radio signal strength based on node positions p_i and p_j (cf. [REKW11]). The dependency between measured and estimated received signal strength can be expressed using distance and radio signal matrices:

$$f(p_i, p_j) = s_{ij} \pm \Delta, \quad \Delta > 0. \tag{4.11}$$

It is hard to exactly specify an upper bound Δ_{max} for the deviation $\Delta = |f(s_{ij}) - d_{ij}|$ of estimated and actual signal strength of two network nodes. The upper bound Δ_{max} should be chosen based on the general setup and dynamically adjusted based on the current situation. For attack recognition, however, an upper bound should be chosen that does not deliver too many false positives and the deviation Δ could be used as measure for the likelihood of suspicion. The larger the deviation Δ the more likely the data of a node i is incorrect.

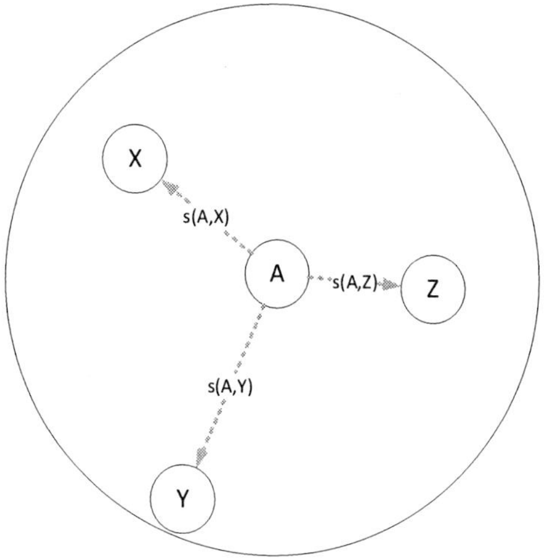

Figure 4.11.: Radio Signal Strength Checking

Furhermore, the relationship between received radio signal strengths of neighbor nodes

$$s_{ij} \leq s_{ik}$$

should be equal to the relationship of the estimated received radio signal strength based on node geo coordinates

$$f(p_i, p_j) \leq f(p_i, p_k).$$

4.4. Summary

In this chapter we presented a concept for cross data analysis in order to increase robustness in situation awareness in tactical MANETs. Firstly, an analysis of data sources available in tactical MANETs for situation awareness was performed: direct and indirect sensor data, mission specific and general knowledge sources. Potential overlaps of these data sources and cross-relationships were examined and categorized. Then inconsistency and consistency of data sources was discussed and a concept for modeling inconsistencies was defined.

The developed concept of cross data analysis was applied to location related data sources in order to provide a proof of concept [ES08]. Location related data sources and their cross-relationships were analyzed and modeled in a consistent way. Radio signal strength was successfully incorporated as indirect sensor data for distance estimation based on a topography based radio propagation model [AE08a, REKW11]. Based on consistency checks of available data set we developed algorithms for the discovery of inconsistencies and the detection of active attacks in tactical MANETs [ES08]. The evaluation of this proof of concept shows that cross data analysis is a powerful approach for robust situation awareness (see chapter 7).

5. Cooperative Trust Assessment

In this chapter we analyze how nodes can cooperate in tactical MANETs in a trusted and robust way and develop cooperative mechanisms that allow the assessment of the trustworthiness of other network nodes, particularly in the case of attacks or node failures. The proposed mechanisms address the specific requirements in tactical MANETs, particularly efficiency, and show how they can be enhanced to incorporate also the trustworthiness of other nodes into indirect trust assessment. For this purpose trust is complemented with a confidence value that reflects the assumed certainty and accuracy of a specific trust estimation.

5.1. Trust Assessment in Tactical MANETs

In the following we review the specific properties and challenges in tactical MANETs with respect to cooperation and trust assessment and derive design objectives for a cooperative trust assessment system.

5.1.1. Specific Properties and Challenges

The properties and constraints of tactical MANETs lead to specific challenges for cooperative trust assessment in such environments. In the following the most important properties of tactical MANET environments (cf. section 2.2.3 on page 27) that are relevant to trust management and the resulting challenges are presented.

There are certain *resource constraints and limitations* in respect to the available communication bandwidth and processing power of mobile devices. Therefore a comprehensive trust management infrastructure, e. g., a public-key infrastructure (PKI) with online validity checks of certificate status, is not suitable for this kind of scenario.

The *network topology changes dynamically and frequently* due to node mobility; new routes are available and existing links may break. Therefore also neighborhood relationships and paths to distant nodes constantly change and no durable, fixed trust evaluation structure can be established. Nodes need to determine the trustworthiness of other nodes and cannot solely rely on a static, pre-established trust infrastructure (e. g., digital certificates) which requires significant overhead and cannot be recovered once compromised.

The MANET consists of a *closed user group*, i. e., participating nodes are known at the start of a specific mission in the targeted application scenario and can be clearly identified and authenticated. Therefore high initial trust is placed on all other network nodes, but may decrease severely if alerts about malicious behavior occur and strong suspicion of potential node compromise arise. These assumptions also hold for coalition environments where trustworthy authentication material is available and high initial trust is placed on coalition nodes.

As all nodes in a tactical MANET are very *determined to achieve the mission objectives and cooperate* utilizing all of their available capabilities high expectations and benchmarks regarding their benign behavior and their willingness for cooperation can be set. On the other side a *rigid policy* is in place that forcefully excludes identified malicious nodes from the network.

Another aspect to be considered is the *expected adversary model* (see section 2.2.4 on page 28) in tactical MANETs. This may also provide some hints and guidelines for optimizing the design of suitable cooperative trust assessment mechanisms. However the proposed concept should generally be able to cope with any kind of attack and open to new threats that may not even be envisioned during the design of the trust assessment system.

It is expected that *attackers are legitimate participants of the network* and exploit their trustworthy position to have a significant impact on the tactical MANET. An attacker may have compromised a regular MANET node and therefore hold all required authentication and encryption keys. Therefore a solution that is based on cryptographic protection mechanisms is not sufficient for these purposes. A cooperative trust assessment is required to effectively detect malicious nodes based on cooperative monitoring of other network nodes and a joint assessment and reaction to detected malicious behavior. There might be even multiple and cooperative attackers that join their efforts potentially exposing Byzantine behavior in order to hide their malicious behavior (cf. section 2.2.4 on page 28).

In tactical MANET *network partitioning* has to be considered as a particular threat as attackers may gain a tactical advantage by disrupting the network performance or isolate a specific group of network nodes. Therefore the trust assessment mechanism has to work efficiently and effectively also in the case that not all network nodes are available but only a limited number of nodes in the neighborhood.

5.1.2. Design Objectives

In the following the design objectives for a cooperative trust assessment system are derived based on the specific properties and challenges in tactical MANETs. The paramount objective is the *precise and accurate assessment of the current trustworthiness of the nodes of the network.* Misbehavior has to be unambiguously detected; a high detection rate should to be achieved while the *number of false-positives has to be as low as possible.*

Besides this general design objective that more or less holds for all trust assessment mechanism a number of specific design objectives have to be considered for tactical MANETs which are described in more detail in the following.

- *Efficiency:* The trust assessment mechanisms have to be light-weight and simple regarding resource consumption. The communicational overhead, i. e., the number of additional packets to be send and the overall amount of data to be exchanged for trust assessment, needs to be minimized.
- *Dezentralized Architecture:* A decentralized architecture is required that allows nodes to cooperatively analyze the current situation, detect malicious behavior and assess the trustworthiness of all other network nodes. The network consists of equal peer nodes and no central instance should serve as a trust anchor which could be a bottleneck or a single point of failure.
- *Robustness against Dynamic Network Changes, Network Fragmentation and Disruptions:* The trust assessment model must not be affected by dynamic network changes, network fragmentation and disruptions. It should be capable to easily

cope with dynamic changes of the network topologies, i. e., changes of neighborhood relationships and paths to distant nodes. The model should not rely on any kind of stable trust or network structure; it should flexibly work with currently available (neighbor) nodes.

- *Dynamic:* The trust assessment system should react if trust estimations provided by other nodes indicate a potential attack. It should dynamically respond to significant changes of the cooperative trustworthiness assessments for certain network nodes. A strong reaction of the trust assessment system is the basis for decisive actions of the overall network to react on malicious network nodes.
- *Robustness against Multiple Attackers:* The cooperative trust assessment model should be resilient to malicious nodes. The trust assessment model should be robust against multiple cooperating malicious nodes. Even if a large number of attackers join their efforts the benign nodes should prevail and detect malicious behavior in order to take adequate response measures.

The overall objective is a fast, robust, efficient, and accurate trust assessment model for the detection of malicious nodes in order to take adequate response measures. As presented in section 3.3.1 there is significant previous work on trust assessment. However, even approaches that were proposed for MANETs are not specifically designed for use in tactical environments considering the specific challenges and design objectives presented above (cf. [CSC11]).

5.2. Efficient Cooperative Trust Assessment

Specifically tailored efficient trust assessment mechanisms are required for tactical MANETs that allow a cooperative trust assessment to provide robust situation awareness. The objective is a shared assessment of the current situation in order to detect threats and attackers (if available). In the following section an overview of the basic concepts and objectives of cooperative trust assessment in MANETs is presented. Subsequently two decentralized assessment architectures – a flooding and a gossiping based approach – are presented and compared with each other.

5.2.1. Overview

The objective of cooperative trust assessment is to locally obtain a trust estimation regarding a specific node which is as close to the aggregation (according to a specific model) of all available trust assessments of other nodes about this node as possible.

5.2.1.1. Notation

The MANET is modeled as a directed graph $G = (V, E)$, where V is the set of vertices (nodes) and E the set of edges (symmetric or asymmetric links). Individual network nodes are denoted with upper case letters and $N = |V|$ is the known and fixed number of mobile nodes. E is a subset of all ordered pairs created by the Cartesian product $V \times V$ of all vertices. WLOG trust values are modeled as real numbers within the interval $[0, 1]$.

- N: known and fixed number of mobile nodes, $N = |V|$
- $N_j(i)$: set of neighbors of node $j \in V$ at time i
- $N_j^*(i) = \{j\} \cup N_j(i)$: set including node $j \in V$ and all neighbors of j at time i
- $t_{kj}^*(i) \in [0, 1]$: trust assessment based on actual local observations about node k at node j $(j, k \in V)$ at time i
- $t_{kj}(i) \in [0, 1]$: current local trust estimation about node k at node j $(j, k \in V)$ at time i (based on available trust assessments)
- $\hat{t}_k(i) \in [0, 1]$: global trust estimation based on combination of all trust assessments that were provided about node $k \in V$ up to time i
- $t_{def} \in [0, 1]$ (e. g., 0.5): default trust value for node k if node j has no trust assessment available (provided as $t_{kj}^*, j, k \in V$)

5.2.1.2. Cooperative Trust Assessment Model

In the following sections we start for simplicity reasons with a simple aggregation model. In this model the trust estimation for a specific network node is based on an estimation of the average of all available trust assessments of other nodes.

Objective The objective is to obtain a trust estimation $t_{kj}(i)$ regarding node k at node j which is as close to the average of all trust assessments at time i as possible:

$$\hat{t}_k(i) = \frac{1}{N} \sum_{l=1}^{N} t^*_{kl}(i)$$

Basic Concept The trustworthiness of neighbor nodes is assessed based on local observations, e. g., consistency checks as described in chapter 4, and data received from adjacent nodes. Based on (a significant number of observations) a trust value t^*_{kj} is determined about node k at node j. These trust assessments are shared with other nodes in order to get a common trust value estimation.

Two communication models for the distribution and exchange of locally assessed trust values are described and compared in the following: a flooding based model which forwards complete trust data records to all other network nodes and a gossiping based trust data dissemination model which is based on the local aggregation and exchange of data values between direct neighbor nodes.

Flooding based Trust Assessment Architecture Trust estimations of neighbor nodes are assessed based on local observations and compiled into a data record. These trust data records are flooded into the network. Intermittent nodes are assumed to not change these data records and forward them until all nodes are reached. This way each node gets a complete view of the network in respect to each network node including all trust data records that were provided by other network nodes. The flooding based architecture is presented in more detail in the following section.

Gossiping based Trust Assessment Architecture Equal nodes contribute to a distributed trust assessment in a peer-to-peer manner. The idea is a gossiping based exchange of opinions about the trustworthiness of other network nodes. "Rumors" (trust and confidence values) about the trustworthiness of network nodes spread from one node to the next node. This way eventually a shared global view on the trustworthiness of each network node is reached which is dynamically updated when new observations lead to changes in trustworthiness estimations. The gossiping based trust assessment architecture is presented in more detail in section 5.2.3.

5.2.2. Flooding based Trust Assessment Architecture

Trust values based on direct observations are flooded to all other network nodes. This way each node receives all trust information that was provided by other network nodes and can correlate all corresponding trust data records for a specific target node.

A flooding based trust assessment architecture for MANETs was, for example, proposed by Miranda and Rodrigues [MR03] where each node periodically floods trust information records about other nodes into the network. Liu *et al.* [LJT04] propose a flooding based mechanism to propagate trust reports containing trust level estimations for other network nodes. In order to limit the network overhead the number of hops that each reports is forwarded is limited to specified upper bound.

In the following the flooding based assessment architecture is described in more detail.

5.2.2.1. Basic Trust Exchange Mechanism

The trust assessment architecture is based on trust values that are locally assessed and then forwarded to the other network nodes.

Local Trust Assessment For each trust assessment t^*_{kj} based on direct observations about node k by a node j a data record is created, e. g., Based on consistency checks as described in chapter 4. This data record contains the target node k, the trust value t^*_{kj} as well as the identity of the locally assessing node j.

Trust Data Forwarding All data records that are created in this way are flooded to all other network nodes. The communication model is a simple flooding mechanism that incorporates duplicate detection. This way all nodes receive all data trust data records that were created by all other network nodes.

Trust Estimation In order to calculate estimations of the trust levels of the other network nodes each node stores all received trust assessments t^*_{kj} records. This includes local assessments as well as contributions provided by other nodes.

Based on the available data records each node j locally calculates an estimation of the trust value of another node k according to the following equation:

$$t_{kj}(i) = \frac{1}{N} \sum_{j=1}^{N} t^*_{kj}(i-1)$$

If no trust assessment is available for node k at node j the default trust value t_{def} is used as $t^*_{kj}(i-1)$.

5.2.2.2. Trust Assessment Updates

When the trust assessment at node l about node k changes at time i, i. e., $t^*_{kl}(i) \neq t^*_{kl}(i-1)$, the new value is simply flooded into the network in order to gradually replace the previously sent trust data record. This is performed in parallel by all nodes which have trust assessment updates available at a specific time i.

5.2.3. Gossiping based Trust Assessment Architecture

In this section the gossiping based trust assessment architecture is described in more detail. The general idea is that trust information is exchanged and aggregated with neighbor nodes in a very efficient ways. No global storing, forwarding (flooding) and evaluation of trust records provided by distant network nodes is required.

The principle of *Gossiping* based communication (also called all-to-all communication) is quite simple: Every node knows a specific item of information which should be communicated to every other node in the network [HHL88]. When a new observation ("rumor") is available at one agent, this agent randomly chooses another agent and forwards it. This procedure is continually repeated in each round by all agent that already know the information ("rumor").

Uniform gossip particularly means that the communication partner is chosen randomly and uniformly among all network nodes [KDG03]. Uniform gossiping protocols have some disadvantages in MANETs. Randomly choosing an arbitrary network node may involve some significant overhead. Routing is expensive in MANETs and randomly selecting a target in each round prevents route reuse [FGR*07].

On the other side the specific characteristics of MANETs may be exploited for gossiping as MANET communication is based on a wireless broadcast medium. Each message that is sent by a node in a specific round may be broadcast and received by all neighbor nodes within the transmission range without any additional cost. Therefore the idea of *broadcast gossip* is that a node does not select specific target nodes but simply broadcasts a message to all nodes within communication range [FGR*07].

Broadcast gossip is the preferred implementation for gossiping based trust assessment in MANETS. In each round each agent that already knows the item of information (trust estimation) forwards it to all direct neighbor nodes. In this gossiping model the receiver is not randomly picked by the sender, instead the group of receivers is determined by the network topology [FKMR09].

5.2.3.1. Basic Trust Exchange Mechanism

As described above the objective is to obtain a trust estimation $t_{kj}(i)$ regarding node k which is as close to the average of all trust assessments at time i as possible. Related work for gossiping based information aggregation [KDG03, BZC11] proposed protocols for the cooperative calculation of average values by network nodes. The Push-Sum Algorithm proposed by Kempe *et al.* [KDG03]) tries to solve this problem in a distributed manner.

In the following a gossiping based solution of the trust assessment problem based on the Push-Sum Algorithm is presented. In order to cooperatively calculate an estimation of the trust value of node k all other network nodes $j \neq k$, $N-1$ perform in parallel the algorithm that is described in following.

Notation

- $a_{kj}(i) \in [0,N]$: current trust estimation aggregate about node k at node j at time i (based on aggregation of trust assessments)

- $w_{kj}(i) \in [0,N]$: current weight assigned to trust estimation aggregate about node k at node j at time i

Local Trust Assessment and Initialization The initial trust estimation $t^*_{kj}(t_0)$ is based on direct observations about node k by a node j (e. g., consistency checks as described in chapter 4):

- trust aggregate: $a_{kj}(t_0) = t^*_{kj}(t_0)$ (default value t_{def} if no assessment yet available)
- weight: $w_{kj}(t_0) = 1$

Trust Data Forwarding and Trust Estimation The following steps are performed in parallel on each node j for each other node k:

1. Let $(\tilde{a}_{kr}(i-1), \tilde{w}_{kr}(i-1))$ be all pairs with trust estimation aggregates about node k sent to node j in round $i-1$
2. Update
$$a_{kj}(i) \leftarrow \sum_r \tilde{a}_{kr}(i-1),$$
$$w_{kj}(i) \leftarrow \sum_r \tilde{w}_{kr}(i-1)$$
3. Choose a non-negative share $\alpha_{kj,l}(i)$ for nodes $l \in N^*_j(i)$, such that $\sum_l \alpha_{kj,l}(i) = 1$ (e. g., $\alpha_{kj,l}(i) = \frac{1}{|N^*_j(i)|}$)
(*Assumption:* Neighbor nodes $N_j(i)$ are known, e. g., based on HELLO message mechanisms as in Optimized Link State Routing (OLSR) [CJ03] or the Ad-hoc On-demand Distance Vector (AODV) [PBRD03] routing protocols.)
4. Send the pair $(\alpha_{kj,l}(i) \cdot a_{kj}(i), \alpha_{kj,l}(i) \cdot w_{kj})$ to each node l
5. $t_{kj}(i) = \frac{a_{kj}(i)}{w_{kj}(i)}$ is the trust estimate about node k of node j in step i

An option would be to exclude trust estimation about neighbor nodes from broadcast messages as these can be heard by all neighbor nodes. This would be a straightforward approach in social environment where human beings exchange rumors about other participants, e. g., in order to not harm their feelings. In a technical system and in particular in a tactical MANET scenario this it not necessary. In contrary this provides transparency and may be used to detect malicious nodes or inconsistencies.

Mass Conservation Property In each round the following properties hold:

$$\sum_{j=1}^{N} w_{kj}(i) = N$$

$$\frac{1}{N} \cdot \sum_{j=1}^{N} a_{kj}(i) = \hat{t}_k(i)$$

This is an important property in order to assure the convergence of the trust estimation values to the correct target value.

5.2.3.2. Trust Assessment Updates

When the trust assessment at node l about node k changes at time i, i. e., $t^*_{kl}(i) \neq t^*_{kl}(i-1)$, the trust estimation for node k is adjusted in the following way:

$$\Delta t^* = t^*_{kl}(i) - t^*_{kl}(i-1)$$

In the following only the case that a single trust assessment update is available at a specific time i is considered but the explanations also apply for multiple trust assessment updates at the same time.

The average trust assessment value before update is (including the mass conservation property):

$$\hat{t}_k(i-1) = \frac{1}{N} \sum_{j=1}^{N} t^*_{kj}(i-1) = \frac{1}{N} \cdot \sum_{j=1}^{N} a_{kj}(i-1)$$

The mass conservation property (last equality in the following) should also hold after the update of the average trust assessment value:

$$\hat{t}_k(i) = \frac{1}{N} \sum_{j=1}^{N} t^*_{kj}(i) \stackrel{!}{=} \frac{1}{N} \cdot \sum_{j=1}^{N} a_{kj}(i)$$

In order to achieve this we adjust $a_{kl}(i)$ in the following way:

$$a_{kl}(i) = a_{kl}(i-1) + \Delta t^*$$

Proof It can be easily shown that the mass conservation property also holds after the update:

$$\frac{1}{N}\cdot\sum_{j=1}^{N} a_{kj}(i) = \frac{1}{N}\cdot\left[a_{kl}(i) + \sum_{j=1, j\neq k}^{N} a_{kj}(i)\right]$$

$$= \frac{1}{N}\cdot\left[(a_{kl}(i-1)+\Delta t^*) + \sum_{j=1, j\neq k}^{N} a_{kj}(i-1)\right] = \frac{1}{N}\cdot\left[\Delta t^* + \sum_{j=1}^{N} a_{kj}(i-1)\right]$$

$$= \frac{1}{N}\cdot\left[(t^*_{kl}(i) - t^*_{kl}(i-1)) + \sum_{j=1}^{N} t^*_{kj}(i-1)\right]$$

$$= \frac{1}{N}\cdot\left[t^*_{kl}(i) - t^*_{kl}(i-1) + t^*_{kl}(i-1) + \sum_{j=1, j\neq k}^{N} t^*_{kj}(i-1)\right]$$

$$= \frac{1}{N}\cdot\left[t^*_{kl}(i) + \sum_{j=1, j\neq k}^{N} t^*_{kj}(i)\right]$$

$$= \frac{1}{N}\cdot\sum_{j=1}^{N} t^*_{kj}(i)$$

5.2.4. Performance Evaluation and Comparison

In the following we evaluate the two communication models – flooding based and gossiping based trust assessment architecture – in respect to the following criteria:

- communication complexity,
- computational complexity,
- storage complexity, and
- accuracy of trust value estimation (convergence and diffusion speed of trust information).

In the following sections each the evaluation for each of the criteria is analyzed individually.

5.2.4.1. Communication Complexity

The communication cost in respect to the number of network nodes N are evaluated for each of the two approaches. Costs are determined as the total amount of data (number of values) that need to be transmitted in each round. For simplicity reasons transmission failures and resulting retransmissions are not explicitly considered. Additionally, it is assumed that trust assessment updates are available at each node in each round and therefore exchanged with the other network nodes.

- *Flooding based approach:* Flooding of trust messages into the network.
  - One message is sent by each node.
    Total number of messages per round: N
  - Each message contains a trust estimation value from each network node for all other network nodes (except sending/receiving nodes)
    Size of each message (number of values): $O(N^2)$

  Number of values that need to be transmitted: $O(N^3)$
  $\Rightarrow$ *Communication cost:* $O(N^3)$
  (Calculation in respect to the number of network nodes N.)
- *Gossiping based approach:* Continuous exchange of trust values with direct neighbor nodes.
  - One message is sent by each node.
    Total number of messages per round: N
  - Each message contains a trust value for the local node for all other network nodes (except sending/receiving nodes)
    Size of each message (number of values): O(N)

  Number of values that need to be transmitted: $O(N^2)$
  $\Rightarrow$ *Communication cost:* $O(N^2)$
  (Calculation in respect to the number of network nodes N.)

The resulting communication complexity for the flooding based trust assessment approach $O(N^3)$ is higher in comparison to the gossiping based approach $O(N^2)$.

5.2.4.2. Computational Complexity

The computational cost in respect to the number of network nodes N are evaluated of each of the two approaches. Costs are determined based on the total number of linear combinations with complexity $O(N)$ that needs to be performed in each round.

- *Flooding based approach:* Calculation of trust estimation for each other network node.
  - Each node has to calculate the linear combination with complexity $O(N)$ of trust values of all $N-1$ other network nodes.
    Total number of linear combinations per round: $N \cdot (N-1)$

  ⇒ *Computational cost:* $O(N^3)$
  (Calculation in respect to the number of network nodes N.)
- *Gossiping based approach:* Calculation of trust estimation for each other network node.
  - Each node has to calculate the linear combination with complexity $O(N)$ of trust values and weights of all $N-1$ other network nodes.
    Total number of linear combinations per round: $2 \cdot N \cdot (N-1)$

  ⇒ *Computational cost:* $O(N^3)$
  (Calculation in respect to the number of network nodes N.)

The resulting computational complexity is in the same order of magnitude for the flooding based trust assessment approach and the gossiping based approach: $O(N^3)$ (represented as required number of linear combinations with complexity $O(N)$ in respect to the number of network nodes N).

5.2.4.3. Storage Complexity

The storage cost in respect to the number of network nodes N are evaluated of each of the two approaches. Costs are determined as the number of trust values that need to be stored.

- *Flooding based approach:* Storage of trust assessments by each network node for each other node.

- Up to $N-1$ trust assessments for each of the $N-2$ other network nodes are stored at each node.
  Total number of stored values: $N \cdot (N-1)(N-2)$

  $\Rightarrow$ *Storage cost:* $O(N^3)$
  (Calculation in respect to the number of network nodes N.)
- *Gossiping based approach:* Storage of trust estimation by each network node for each network node.
  - $N-1$ times two values (trust aggregates and weights) are stored at each node.
    Total number of stored values: $2 \cdot N \cdot (N-1)$

  $\Rightarrow$ *Storage cost:* $O(N^2)$
  (Calculation in respect to the number of network nodes N.)

The resulting storage cost is higher for the flooding based trust assessment approach $O(N^3)$ in comparison to the gossiping based approach $O(N^2)$ (Represented as number values that need to be stored in respect to the number of network nodes N).

5.2.4.4. Estimation Accuracy: Convergence and Diffusion Speed

In this section the two approaches are evaluated in respect to their estimation accuracy. For this purpose the convergence of the trust estimation process and the diffusion speed of trust information are analyzed.

For the evaluation of the convergence of the trust estimation process it is assumed that the trust assessments do not change (at least for the specific limited period of time of consideration). For stable network situations this may be a reasonable assumption as trust assessments will not or only marginally change. This is of course not be expected in a dynamic tactical MANET but allows to specify the reached convergence level in a rational and simple way.

A more sophisticated analysis may also consider the changes of trust assessments for specific network nodes. In highly dynamic scenarios with constant changes of trust assessment values the estimation process may not converge at all but the analysis presented in the following may still deliver important insights in the basic characteristics and the the accuracy of the trust estimation process.

Based on [KDG03] we define the diffusion speed of trust information. For this purpose we introduce a contribution vector $\mathbf{v}_j(i)$. Each component of the vector reflects the contribution from a specific network node to the current trust estimation at node j. Perfect convergence of the trust estimations is achieved if all components reach the value $\frac{1}{N}$. We define the relative error of the trust estimation at node j at time i to be $\Delta_j(i) = \max_k \left| \frac{\mathbf{v}_{j,k}(i)}{\|\mathbf{v}_j(i)\|_1} - \frac{1}{N} \right| = \left\| \frac{\mathbf{v}_j(i)}{\|\mathbf{v}_j(i)\|_1} - \frac{1}{N} \cdot \mathbf{1} \right\|_\infty$.

Definition 5.1. *We say that* $T = T(\delta, N, \varepsilon)$ *is (an upper bound on) the* diffusion speed *of the trust assessment mechanism if* $\max \Delta_j(i) \leq \varepsilon$ *with probability at least* δ *at all times* $i \geq T(\delta, N, \varepsilon)$.

Convergence and Diffusion Speed of Flooding based Approach The flooding based approach converges at the latest when all trust assessment values have been forwarded to each network node. Trust messages are further forwarded in each round, therefore the trust estimation process progresses step by step and will eventually converge in a finite network.

As trust messages propagate one hop per round the network diameter is an upper limit for the maximum number of rounds that are required in order to reach full convergence. The maximum network diameter is reached for a linear setup where all nodes form a chain. Therefore the upper bound for the network diameter is $N-1$. Hence the number of steps for the completion of the convergence process is $O(N)$. Therefore also the diffusion speed $T(\delta, N, \varepsilon) = O(N)$.

Convergence and Diffusion Speed of Gossiping based Approach The diffusion speed for the gossiping based approach is hard to model for general case as it depends very much on node mobility and network topology. In a dense network where each node has a lot of neighbors the diffusion speed is a lot higher in comparison to sparse and spread out network.

Kempe *et al.* present [KDG03] present a theoretic model and some upper bounds for the diffusion speed of gossiping based information aggregation. Their analysis of the diffusion speed of the gossiping based approach is based on the assumption that the shares $\alpha_{kj,l}$ that are exchange with neighbor nodes can be modeled as a matrix

$A = (\alpha_{jl})_{jl}$, where the entry α_{jl} denotes what fraction node j sends to l (in the trust estimation process for node k). They leverage related work on the mixing speed of Markov Chains by examining the probability that the Markov Chain defined by A (and starting at node l) is at states j at time t. The following theorem is derived based on this analysis for the diffusion speed of the gossiping based approach [KDG03]:

Theorem Let T' be a function such that $max_k||\frac{\mathbf{e}_k^T \cdot A^t}{\pi} - 1||_2 \leq \varepsilon$, for all i $\geq T'(N,\varepsilon)$. Then, an upper bound on the diffusion speed for the flooding mechanism defined by A is given by $T(N,\varepsilon) := 2T'(N, \sqrt{\frac{N\varepsilon}{2+N\varepsilon}})$.

This theorem enables leveraging substantial related work on the convergence speed of Markov Chains and Random Walks. For example, if the underlying network is an expander the diffusion speed $T(N,\varepsilon) = O(\log N + \log\frac{1}{\varepsilon})$ [KDG03].

5.2.4.5. Summary of Evaluation Results

Table 5.1 shows a summary for the first three evaluation criteria: communication, computational and storage complexity The analysis of the convergence and diffusion speed is missing as this analysis is more complex as described in the section above. The diffusion speed of the flooding based approach is $O(N)$. The diffusion speed of the gossiping based approach, however, depends very much on node mobility and network topology but is generally expected to be lower than for the flooding based approach. For example, if the underlying network is an expander the diffusion speed of the gossiping approach is $O(\log N + \log\frac{1}{\varepsilon})$.

Table 5.1.: Evaluation of Trust Assessment Architectures

Evaluation Criteria	**Flooding based Approach**	**Gossiping based Approach**
Communication Complexity	$O(N^3)$	$O(N^2)$
Computational Complexity	$O(N^3)$	$O(N^3)$
Storage Complexity	$O(N^3)$	$O(N^2)$

The computational complexity is similar for both approaches: $O(N^3)$. The gossiping based trust assessment model is more efficient in respect to the communication complexity: $O(N^2)$ in comparison to $O(N^3)$ for the flooding based approach. Additionally also the storage complexity is also lower: $O(N^2)$ in comparison to $O(N^3)$ for flooding based approach.

In conclusion the gossiping based communication model provides an efficient basis for robust cooperative trust assessment in tactical MANETs. However, this basic mechanism does not provide any protection against faulty trust information provided by faulty, misconfigured, selfish or malicious nodes.

5.3. Enhanced Gossiping based Trust Assessment Architecture

In this section the efficient gossiping based trust assessment architecture is extended in order to increase robustness in respect to selfish and malicious nodes. Nodes may provide faulty trust information when they are compromised by an attacker or due to misconfiguration or failures. Since the processed and aggregated information is not some arbitrary sensor measurements but information reflecting the trustworthiness of other network nodes this information may be used to enhance the data aggregation process. The objective is to mitigate the influence of spurious ratings based on enhanced trust aggregation mechanisms.

5.3.1. General Idea

The general idea is to consider the trustworthiness of the data providing nodes during the aggregation process in order to reduce the influence of faulty or malicious nodes that provide wrong trust information (cf. [BMB08, LY03]). The processed trust values themselves are used for optimizing the aggregation algorithm. This is an important aspect when incorporating second hand information: *Higher weights are assigned to information provided by neighbors with higher trust.*

The trust for node k at node j is calculated as the weighted sum of its own trust value and the trust values of the neighbor nodes r about k [LY03]:

$$t_{kj} = \lambda \cdot \sum_r f(t_{rj}) \cdot t_{kr}$$

where f is a mapping function and λ is a normalization factor.

The function f maps the trustworthiness of neighbor nodes into a weighting of the provided trust information. The weight may be proportional to the trustworthiness or more general a monotonically increasing function. The higher the trust t_{rj}, the higher the weight for the trust value t_{kr} provided by neighbors r.

5.3.2. Weighted Combination of Trust Values

In order to integrate the weighting into the aggregation process the update step has to be adjusted. The received trust estimation values that were sent to node j in a specific round $i-1$ are weighted using a mapping function f and a normalized using a factor λ:

$$a'_{kr}(i-1) \leftarrow \lambda_k(i-1) f(t_{rj}(i-1)) \cdot \tilde{a}_{kr}(i-1)$$

$$w'_{kr}(i-1) \leftarrow \lambda_k(i-1) f(t_{rj}(i-1)) \cdot \tilde{w}_{kr}(i-1)$$

The following factor is applied to normalize the overall sum of all weights :

$$\lambda_k(i-1) = \frac{\sum_{r \in M} \tilde{w}_{kr}(i-1)}{\sum_{r \in M} f(t_{rj}(i-1)) \tilde{w}_{kr}(i-1)}$$

Derivation of λ In order to keep the same overall sum of weights for the enhance trust aggregation model the following equation should hold:

$$\sum_{r \in M} w'_{kr}(i-1) = \lambda_k(i-1) \sum_{r \in M} f(t_{rj}(i-1)) \tilde{w}_{kr}(i-1) \stackrel{!}{=} \sum_{r \in M} \tilde{w}_{kr}(i-1)$$

$$\Rightarrow \lambda_k(i-1) = \frac{\sum_{r \in M} \tilde{w}_{kr}(i-1)}{\sum_{r \in M} f(t_{rj}(i-1)) \tilde{w}_{kr}(i-1)}$$

5.3.3. Enhanced Trust Assessment Model

The local trust assessment, initialization and trust data forwarding are performed in the same way as before. The trust value estimation process is extended by introducing the weighting step.

Trust Value Estimation The following steps are performed in parallel on each node j for each other node k:

1. Let M denote the set of nodes that sent trust estimation aggregates about node k to node j in round $i-1$ and $(a_{kr}(i-1), w_{kr}(i-1))$ the respective value pairs. Then the weighted value pairs are calculated as described above:

$$\tilde{a}_{kr}(i-1) \leftarrow \lambda_k(i-1) f(t_{rj}(i-1)) \cdot a_{kr}(i-1)$$

$$\tilde{w}_{kr}(i-1) \leftarrow \lambda_k(i-1) f(t_{rj}(i-1)) \cdot w_{kr}(i-1)$$

2. Update:

$$a_{kj}(i) \leftarrow \sum_{r \in M} \tilde{a}_{kr}(i-1) = \lambda_k(i-1) \sum_{r \in M} f(t_{rj}(i-1)) \cdot a_{kr}(i-1),$$

$$w_{kj}(i) \leftarrow \sum_{r \in M} \tilde{w}_{kr}(i-1) = \lambda_k(i-1) \sum_{r \in M} f(t_{rj}(i-1)) \cdot w_{kr}(i-1)$$

3. Choose a non-negative share $\alpha_{kj,l}(i)$ for nodes $l \in N_j^*(i)$, such that

$$\sum_l \alpha_{kj,l}(i) = 1 \qquad (\text{e. g., } \alpha_{kj,l}(i) = \tfrac{1}{|N_j^*(i)|})$$

4. Send the pair $(\alpha_{jk,l}(i) \cdot a_{kj}(i), \alpha_{jk,l}(i) \cdot w_{jk})$ to each node l
5. $t_{kj}(i) = \frac{a_{kj}(i)}{w_{kj}(i)}$ is the trust estimate about node k of node j in step i

Mass Conservation Property In each round the following property still holds:

$$\sum_j w_{kj}(i) = N$$

The second property does not hold in this case anymore as trust assessment values are intentionally not weighted equally in the enhanced trust assessment architecture: more trusted nodes contribute more to the overall trust estimate. Therefore the target trust estimation is not the equally weighted average of all trust assessments.

Trust Assessment Updates When the trust assessment value at node l about node k changes at time i, i. e., $t^*_{kl}(i) \neq t^*_{kl}(i-1)$, the trust estimation for node k is adjusted in the same way as before:

$$\Delta t^* = t^*_{kl}(i) - t^*_{kl}(i-1)$$

The enhancement of the trust aggregation mechanism presented in this section reduces the influence of faulty or malicious nodes that provide wrong trust information. This approach is further elaborated and extended in the next section and a trust assessment system based on this architecture is presented.

5.4. TEREC

An instantiation of an efficient cooperative trust assessment model for tactical MANETs based on the gossiping based architecture described above is presented in the following.

The model was developed within the Master thesis supervised by me of Norbert Bißmeyer [Biß08]. The results were further elaborated as the *Trust Evaluation and Rumor based Exchange for Cooperative Trust Assessment (TEREC)* system and published in a joint paper [EB09]. This system incorporates the presented trust assessment mechanisms which are further extended and tailored in order to address the specific requirements in tactical MANETs. In particular, the model of trust is extended and complemented with a measure of the reliability of a specific trust estimation (confidence).

5.4.1. Overview

We propose a cooperative approach for trust assessment in tactical MANETS. The core is a decentralized gossiping based trust assessment framework that consists of a network of cooperating peer nodes that exchange trust information. Equal nodes contribute to a distributed trust assessment in a peer-to-peer manner. There are no hubs or central nodes that have a distinguished position or role.

Each node does a local assessment of direct neighbor nodes and shares the results in a cooperative way with all other network nodes. "Rumors" (trust and confidence values) about the trustworthiness of network nodes spread from one node to the next node. There is no flooding mechanism that would forward the assessment results into the network or complicated broadcasting mechanism that would require management mechanisms and forwarding selection procedures and message transmissions. Nevertheless a shared global view on the trustworthiness of each network node is reached.

The following gives an overview on four important building blocks of the gossiping based TEREC architecture: *Trust Modeling and Assessment*, *Gossiping based Trust Assessment Architecture*, *Trust Combination* and *Trust Assessment Protocol*.

Trust Modeling and Trust Assessment The trustworthiness of other network nodes is modeled as a (trust, confidence) value pair where trust represents an estimation for the trustworthiness of a specific node and confidence models the certainty of this statement. The trustworthiness is assessed based on local observations, e. g., consistency checks, and data received from adjacent nodes. Trust and confidence values are derived from the data gathered based on the beta distribution. The local trust assessment is described in more detail in section 5.4.2.

Trust Combination Opinions (trust and confidence values) received from other network nodes need to be combined. This is particularly important for the assessment of a distant node where the goal is an estimation of the trustworthiness of this individual node and not the trustworthiness of a path to this node. There are two specific aspects that have to be considered:

- *Indirect Assessment (Serial Combination):* Opinions received by a node about a third node should be incorporated based on the trustworthiness of the providing node.
- *Multiple Opinions (Parallel Combination):* When multiple opinions are received about a specific network node these estimation need to be combined.

The trust combination mechanisms and the related equations for the calculation of trust and confidence values are described in more detail in section 5.4.4.

Trust Assessment Protocol The trust assessment protocol specifies how the components of the trust assessment system on the individual nodes interact and cooperate. The distributed architecture and the systems dynamics have to be considered in order to create a defined process for cooperative trust assessment. The trustworthiness of all network nodes needs to estimated based on the gathered data and and dynamically updated when new information is available. In a round based approach opinions (trust and confidence values) are combined based on the specified calculation procedures. Node distances (hop count) to target nodes are used to decide which opinions are considered for an assessment and which opinions are left out. The trust assessment protocol is presented in more detail in section 5.4.5.

The following sections describe each of these building blocks in more detail.

5.4.2. Trust Assessment and Modeling

The trustworthiness of neighbor nodes is assessed based on local observations, e. g., consistency checks, and data received from adjacent nodes. The results of this local trust assessment serve as basis for the cooperative trust assessment [EB09]. The local trust assessment consist of two step (as shown in figure 5.1): the evaluation of local observations and probabilistic modeling using the beta distribution and the calculation of trust and confidence values based on this assessment.

A common approach for trust modeling is to use the beta distribution in combination with a Bayesian approach (cf. section 3.3.2 on page 81). This way trust and confidence

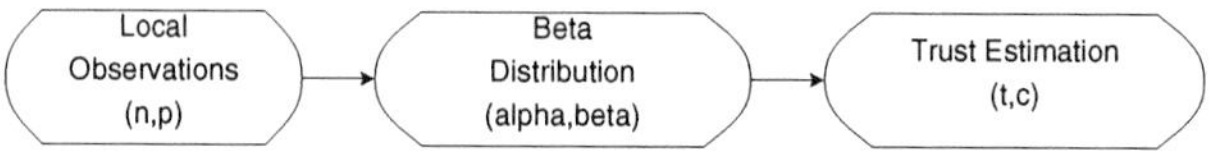

Figure 5.1.: Local Trust Assessment

values can be derived from the parameters of the beta distribution based on the data gathered.

Definition 5.2. *The number of correct (positive) observations within time interval i is denoted as $p(i)$ and the number of malicious (negative) observations as $n(i)$.*

The posterior distribution is updated based on the newly available observations, typically derived as $\alpha(i) = \alpha(i-1) + p(i-1)$, $\beta(i) = \beta(i-1) + n(i-1)$ [BLB03, BB03b]. However, as the influence on the calculation of new observation gets lower over time when the number of observation increases, we use a modified Bayesian approach with an aging factor u (cf. [Buc04, BB04, BMB08]):

$$\alpha(i) = u \cdot \alpha(i-1) + p(i-1) \tag{5.1}$$

$$\beta(i) = u \cdot \beta(i-1) + n(i-1) \tag{5.2}$$

Trust Modeling based on Beta Distribution The trustworthiness of other network nodes is represented by a (trust, confidence) value pair where trust represents an estimation for the trustworthiness of a specific node and confidence models the certainty of this statement (cf. [TB04, TB06]):

Definition 5.3. *The* trust *that node $j \in V$ has regarding node $k \in V$ at time i is denoted as $t_{kj}(i) \in \mathbb{R}$.*

Definition 5.4. *The* confidence *that node j has regarding this estimation is denoted as $c_{kj}(i) \in \mathbb{R}$.*

In the proposed cooperative trust assessment system a modified Bayesian approach is used. The mean of the Beta calculation is used as trust value and the variance is used as the confidence value (as proposed in [ZMHT05, ZMHT06]):

Definition 5.5. *The* trust value *of node $j \in V$ for node $k \in V$ at time i is modeled as the expected value of the Beta distribution:*

$$t_{kj}(i) = \frac{\alpha_{kj}(i)}{\alpha_{kj}(i) + \beta_{kj}(i)} \tag{5.3}$$

Definition 5.6. *The* confidence *value of node j for this trust estimation is modeled as the standard deviation of the Beta distribution normalized to the interval $[0,1]$:*

$$c_{kj}(i) = 1 - \sqrt{12} \cdot \sigma(\alpha_{kj}(i), \beta_{kj}(i)) \tag{5.4}$$

Trust and confidence values are updated at each observation. Furthermore these two values may be combined if a single value representation as *trustworthiness* of node is required, e. g., for application purposes (cf. [ZMHT05]).

5.4.3. Gossiping based Trust Assessment Architecture

As presented above a gossiping based communication model on based on trust information exchange with direct neighbor nodes enables an *efficient trust assessment mechanism* that is also *robust against network fragmentation and disruptions.* This way neither a central entity is required nor the availability of routes to target nodes of the trustworthiness evaluation. "Rumors" (trust and confidence values) about the trustworthiness of network nodes spread from one node to the next node. This way eventually a shared global view on the trustworthiness of each network node is reached which is dynamically updated when new observations lead to changes in trustworthiness estimations.

Each node locally monitors its directly connected neighbor nodes in order to identify spurious or malicious behavior. Based on these observations it generates corresponding trust and confidence values. The resulting trustworthiness information is then exchanged cooperatively with other network nodes.

The local trustworthiness estimates are combined with values provided by neighbor nodes. This way trust values are merged into a combined (trust, confidence) table that represents the overall assessment of the trustworthiness of other nodes. The calculation procedures are described in more detail in the next section.

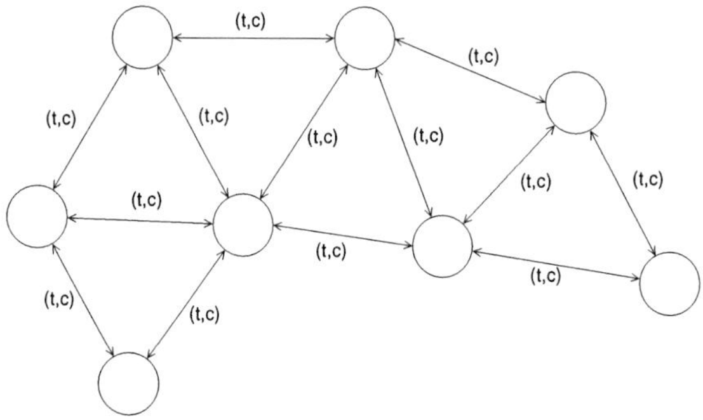

Figure 5.2.: Gossiping based Cooperative Trust Assessment

5.4.4. Trust Value Combination

The goal of this work is to evaluate a network node for trust assessment independent of any specific route or path, therefore mechanisms are required to combine local observations with observation of other network nodes. Previously proposed approaches often exclusively evaluate the trustworthiness of a particular path to a network node for reliable data packet delivery (cf. [ZMHT05, TB04]).

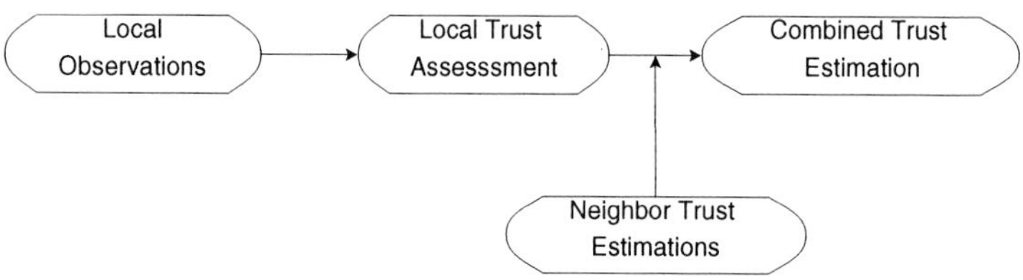

Figure 5.3.: Combination of Local Trust Assessments and Neighbor Trust Estimations for Combined Trust Estimation

Observation of trusted direct neighbor nodes should be weighted as more significant than information provided by distant nodes. Untrusted nodes should contribute to the calculation of node trust with a default trust value for the destination node as no accurate statement can be issued. The default value pair $(t_{def}, c_{def}) = (0.5, 0)$ represents ignorance about the trustworthiness of a node.

There are two specific aspects that have to be considered when combining opinions (trust and confidence values) received from other network nodes:

- *Indirect Assessment (Serial Combination):* Opinions received by a node about a third node should be incorporated based on the trustworthiness of the providing node.
- *Multiple Opinions (Parallel Combination):* When multiple opinions are received about a specific network node these estimation need to be combined.

The following two sections described the trust calculation procedures for both cases in more detail.

5.4.4.1. Indirect Assessment

The combination of trustworthiness values for nodes that are not direct neighbor nodes is challenging. The trustworthiness of a node is estimated by means of trust and confidence values of an adjacent node which provides corresponding trust and confidence values for the selected node (serial combination). The incorporation of opinions received by a node about a third node should be based on the trustworthiness of the providing node. The first calculation provides a mechanism for indirect trust assessment for distant nodes that may be several hops away as shown in figure 5.4.

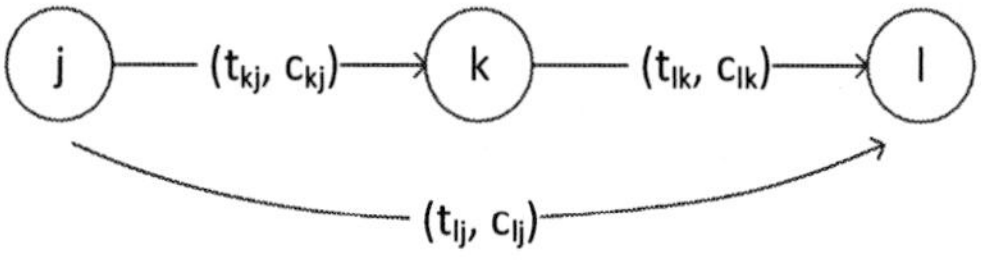

Figure 5.4.: Trust Calculation for Distant Nodes

The function $t_{lj}^{k}(t_{kj}(i), t_{lk}(i), c_{lk}(i))$ with nodes $j,k,l \in V$ is used (if $t_{kj}(i) \geq t_{def}$) to calculate the trust value for node l at node j with trust information from node k about node l at time i where $0 \leq t_{lj}(i) \leq 1$. Furthermore weights $\lambda_{tc}, \lambda_{tt} \in \mathbb{R}$ are used in order to adjust the calculation to be more optimistic with weight values $\lambda < 1$ or pessimistic with weight values $\lambda > 1$.

$$t_{lj}^{k}(i) = t_{def} + (t_{kj}(i) - t_{def})(t_{lk}^{\lambda_{tt}}(i) \cdot c_{lk}^{\lambda_{tc}}(i)) \tag{5.5}$$

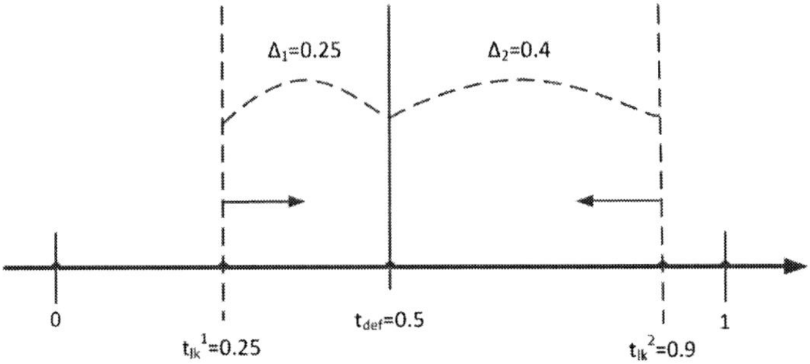

Figure 5.5.: Calculation of Trust Value for Distant Nodes Example

If a node providing trust information does not have a trustworthiness value meeting minimum standards of trust, information regarding a third node cannot be used with confidence and subsequently this third node is assigned a default value for trust and confidence.

As shown in both situations of figure 5.5 second hand trust calculations depend upon appropriate shifting of trust to default values. The trust value $t_{lk}(i)$ is incorporated as long as the trust value for the adjacent node $t_{kj}(i)$ is higher than the default trust value.

In this instance the difference between the default trust and the trust $t_{kj}(i)$ is calculated and multiplied by the product of the trust $t_{lk}(i)$ and confidence $c_{lk}(i)$. Figure 5.5 identifies two independent situations (t^1_{lk} with Δ_1 and t^2_{lk} with Δ_2) where the delta represents the difference in equation (5.5). Depending on trust and confidence values from adjacent node k, trust t^1_{lk} or t^2_{lk} is adjusted towards the default value. The smaller the product of trust $t_{lk}(i)$ and confidence $c_{lk}(i)$, the more the difference values Δ_1 and Δ_2 are scaled down in equation (5.5).

The confidence $c_{kj}(i)$ is not involved in this calculation as the result should give only an aggregate value of trust, not the accuracy of this value. Uncertainty of accuracy for the trust value of node k can be obtained from confidence values calculated in equation (5.6).

The function $c^k_{jl}(t_{kj}(i), c_{kj}(i), c_{lk}(i))$ with nodes $j,k,l \in V$ is used to calculate the confidence in the trust value about node l at node k with information from node k about node l (see equation (5.6)). The weights $\lambda_{cc}, \lambda_{ct} \in \mathbb{R}$ determine the amount of decrease to trust and confidence values respectively.

$$c^k_{jl}(i) = c_{lk}(i) \cdot c^{\lambda_{cc}}_{kj}(i) \cdot t^{\lambda_{ct}}_{kj}(i) \tag{5.6}$$

In contrast to the trust calculation in equation (5.5) confidence values are required to be more dependent on the trust and confidence values of adjacent node k, therefore the equation for confidence is built by multiplication as opposed to addition. In general confidence $c^k_{jl}(i)$ is lower than confidence $c_{lk}(i)$.

Trust $t_{lk}(i)$ is not involved because the amount of trust is not relevant in this calculation. As trust and confidence values are required to be mutually associated in order to correctly calculate trustworthiness, missing trust values can be taken from trust $t^k_{lj}(i)$ where $t_{lk}(i)$ is involved.

5.4.4.2. Multiple Opinions

When multiple opinions are received for a specific network node these estimation need to be combined (parallel combination). The second calculation procedure provides the means for a combination of trust values from multiple neighbor nodes that may have different trust and confidence values for distant nodes as shown in figure 5.6.

Trust $t^{k_t}_{jl}(i)$ and confidence $c^{k_t}_{jl}(i)$ for node k at time i are calculated at node j where node l can be reached by various neighbors $k_t \in V$.

$$t_{lj}(i) = \frac{\sum_{k_t \in N_j(i)} t^{k_t}_{lj}(i) \cdot c^{k_t}_{lj}(i)}{\sum_{k_t \in N_j(i)} c^{k_t}_{lj}(i)} \tag{5.7}$$

For the trust assessment in tactical environments also deviating opinions are considered but the respective confidence value is adjusted accordingly (cf. [BMB08]). This way all newly available knowledge is integrated in order to increase the accuracy of the trust assessment or decrease the confidence in case of deviating evidence.

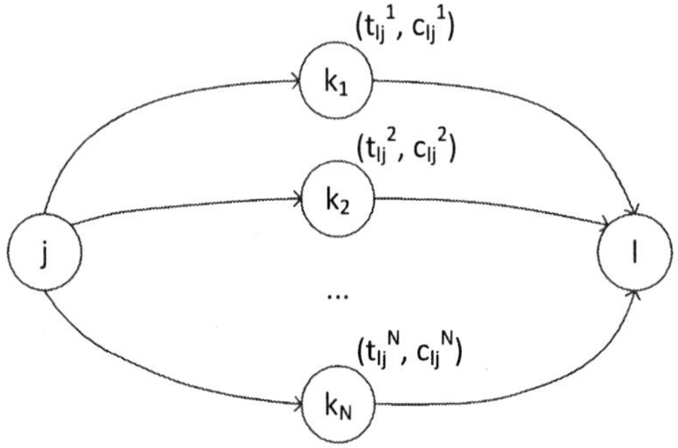

Figure 5.6.: Trust Calculation based on Multiple Estimations

In equation (5.7) trust values derived from several trust values of different neighbors are combined. In the formula the trust assessment of each neighbor node is multiplied by the confidence value assigned to this value and the sum of these values is divided by the sum of the confidence values of all assessments. This equation provides normalized results incorporating the trust assessment of all neighbor nodes and therefore no further weights are necessary. In this equation the assessments of multiple neighbors can be combined into one calculation step where the variable $N_j(i)$ represents the number of neighbors of node j.

Equation (5.8) shows the confidence of a node j calculated from a combination of values provided by several neighbors. In this case it should be checked whether information provided by multiple parties conforms. The congruence of views can be used as basis for the confidence that should be put into the overall trust estimation. This approach is more fine-grained and adaptive than only considering the trustworthiness of the node providing trust information [BMB08]. The proposed calculation ensures that the confidence value is high if the neighbor nodes agree (i. e., the gap between trust values is small) and the reverse if the opinions differ a lot (i. e., trust value differentials are high).

$$c_{lj}(i) = \left(1 - \frac{\sum_{k_s,k_t \in N_j(i), s \neq t} |t_{lj}^{k_s}(i) - t_{lj}^{k_t}(i)|}{|N_j(i)| \cdot (|N_j(i)| - 1)}\right) \sum_{k_t \in N_j(i)} c_{lj}^{k_t}(i) \tag{5.8}$$

The fraction expresses the mean value of the differences between all neighbor nodes.

5.4.5. Trust Assessment Protocol

Trust values are updated in fixed intervals adhering to previous definitions in order to conserve resources at mobile nodes.

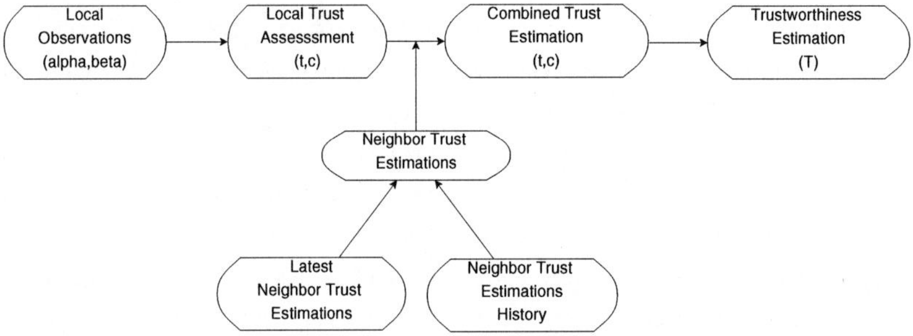

Figure 5.7.: Chain of Calculations in order to Estimate the Trustworthiness of a Node

In figure 5.7 the chain of calculations is shown which is performed on each node in parallel for each other node in order to determine the trustworthiness of other network nodes:

1. *Local Trust Assessment* Local calculation of trust and confidence values (t,c) incorporating positive and negative observations (p,n) is performed based on Bayesian estimation and Beta distribution.
2. *Neighbor Trust Estimations* Second hand trust and confidence values of other nodes are stored in a trust estimation history buffer. Only one (trust, confidence) value pair is evaluated and incorporated for each other node in a specific round of updates and added to the history buffer.

An aging mechanism is applied to the trust estimation values provided by neighbor nodes in order to place higher importance to more recent trustworthiness data. Aging is applied to both trust and confidence values based on an observation window mechanism. A sliding window mechanism and a weighted average of W trust values from the past (as described by [ZMHT05]) is used in order to age older values and incorporate trust assessment values weighted according to their freshness:

$$t_{kj}(i) = \frac{2}{W(W+1)} \sum_{l=1}^{W} l \cdot t_{kj}(i - W + l - 1) \tag{5.9}$$

Only observations within the last W time intervals are involved in the calculation of trust and confidence and the weight of older values is lower than the most current one. This way increased relevance is placed on new observations in comparison to older values and the *trust assessment dynamically responds* to significant changes of trust values for certain network nodes.

Values that were provided by one-hop neighbors with overly high hop counts above a certain threshold Δ_{hop_count} to the destination node are discarded and not used for the trust estimation (see details in section 7.3 on page 246).

3. *Combined Trust Estimations* Trust and confidence values for distant nodes are calculated using equation (5.5) and (5.6) respectively. In those cases where more than one neighbor provides an trust estimation a combination of all values provided is performed using equation (5.7) and (5.8).

The resulting method is a very robust system as evaluated in simulation environments with variable number of attackers (see section 7.3 on page 246 for details). Evaluation results show that the proposed trust assessment scheme allows a benign majority of nodes (in the presence of an increasing amount of malicious nodes) to prevail and that they able to accurately classify network nodes based on trust estimations.

5.5. Summary

We developed efficient and robust mechanisms that allow the cooperative trust assessment of network nodes in the case of attacks or failures [EB09]. Based on a local trust assessment of direct neighbor nodes (e. g., based on detection of routing anomalies results are exchanged with other nodes. The proposed mechanisms are specifically tailored for the characteristics of tactical MANETs. The gossiping based trust assessment architecture exploits the wireless medium by broadcast trust information to all direct neighbor nodes. This enables efficient and simple information aggregation mechanisms.

The effectiveness of the trust assessment mechanisms is further enhanced considering the trust level of the providing node when incorporating second hand trust information. The trust model is complemented with a confidence value representing the certainty about a specific trust assessment. This fosters the mitigation of spurious ratings by checking the congruence of views and lowering the confidence when incorporating contradiction assessments.

Finally, we developed the TEREC system which is a specific instantiation of the proposed mechanisms. For this instantiation the procedures and protocols are described and specified in detail. TEREC has been implemented and evaluated. The results show that the system is very robust system and that trust assessments quickly converge to the correct value even in environments with several attackers. The proposed trust assessment scheme allows a benign majority of nodes to prevail and taccurately classify network nodes based on observations and trust estimations. The evaluation results are presented in more detail in section 7.3 on page 246.

6. Probabilistic State Modeling

In this chapter we present concepts for probabilistic state modeling and estimation and discuss how they can be utilized in order to make situation awareness in tactical MANETs more robust in respect to inaccurate and error-prone input data. Probabilistic state modeling provides higher flexibility and robustness as observations and states can be considered as "probability cloud" instead of specific and explicit values. The probabilistic state estimation process allows incorporating multiple observations and additional knowledge sources in a probabilistic manner and to adjust the likelihood of specific system states accordingly. A specific challenge is the distributed nature of tactical MANETs. We address this challenge by presenting a distributed state estimation architecture resolving temporal and causal correlation of observations and estimations performed by other network nodes using re-simulation. Finally, we present Task Force Tracking (TFT) as an example application which shows how these concepts can be implemented for a specific application scenario.

6.1. Probabilistic State Modeling and Estimation

In this section we present the advantages of probabilistic state estimation for situation awareness in tactical MANETs. We discuss the requirements for probabilistic state modeling in tactical MANETs. Finally we describe how particle filters as the preferred estimation mechanisms as they provide the required capabilities.

6.1.1. State Modeling

Multiple sensors on each network node can be used to provide coordinated measurements of different node and network properties. Based on these measurements an estimation of the current network state can be developed (cf. section 3.4.1 on page 92 for more details). Some of these sensors deliver continuous measurement values and other discrete sets of information. Table 6.1 presents our notation for probabilistic state modeling and estimation.

We assumed that each node has a good knowledge about its own state and its actions. Since mission objectives are shared and nodes communicate and cooperate in tactical MANETs within a closed group of participants (see section 2.2.3 on page 27 for more details) the state of other nodes can be estimated with relatively high accuracy. A consistent modeling of all available dynamic measurements (e.g., node positions, routing protocol state, battery status, application data or additional knowledge sources) and explicit knowledge (e.g., mission profiles and mobility models) is required for this purpose.

Table 6.1.: Notation for Probabilistic State Modeling and Estimation

Symbol	Description
N	The *number of nodes* in the MANET is (for simplicity) assumed to be fixed.
$\mathbf{x}_k^i \in \mathbb{R}^{n_x}$	The *current state* of node i at time t_k . The state is (at least partially) not directly observable but can be inferred from sensor data.
$\mathbf{z}_k^i \in \mathbb{R}^{n_z}$	The *measurements (related to target state)* at time t_k by node i.
$\mathbf{z}_{1:k}^i = \{\mathbf{z}_t^i, t = 1, \ldots, k\}$	The sequence of all available measurements up to time t_k by node i.
$p(\mathbf{x}_k^i \vert \mathbf{z}_{1:k}^i)$	The *posterior probability density function of a node i* regarding all available measurements up to time t_k is also known as filtering distribution. For better readability the node index i is omitted in the following when it is clear from the context to which node a measurement or a state variable refers.

Modeling of Discrete and Continuous State Information The relevant information that describes the situation in a tactical environment can be modeled as discrete and continuous states. Some examples for state information in tactical MANET scenarios are shown in table 6.2 and presented in the following.

- The tracking of team members and available resources engaged in disaster recovery mission provide the basis for situation awareness [Cou05, SG11]. This includes the status of resources and available material, e. g., battery status of devices, remaining vehicle fuel or relief supplies. Additionally information re-

Table 6.2.: Examples of State Information $\mathbf{x}_k^i$ in Tactical MANETs

Feature	Type	Description
Position of entities	Continuous	Node position, e. g., rescue team members, disaster or enemies
Velocity of entities	Continuous	Node movement velocity, e. g., rescue team members, disaster or enemies
Status of resources	Continuous	Resources, e. g., battery status of devices or remaining fuel in vehicles, as well as material, e. g., Relief supplies, e. g., battery status of devices or remaining vehicle fuel
Radio signal strength	Continuous	Received radio signal strength of neighbor node, e.g., Received Signal Strength Indicator (RSSI)
Link states	Discrete	State of local links to neighbor nodes
Network topology	Discrete	Known routes between network nodes
Status of resources	Discrete	Number of available resources, e. g., relief supplies
Vehicle type	Discrete	Type of tracked vehicle, e. g., truck, boat

lated to the disaster is critical to achieve the mission objectives. Victims need to be identified, their health status has to be captured and their location should be tracked. This allows for a matching of detected vs. missing people.

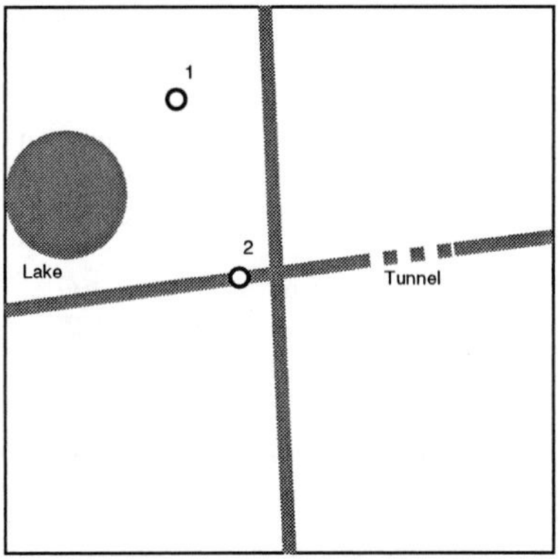

Figure 6.1.: Example for Continuous State Modeling: Location Tracking

The tracking of location and velocity of entities is a typical example of *continuous state* information. Figure 6.1 shows an example of location tracking in tactical

MANETs with two mobile nodes representing two vehicles: one moving in the open field close to a lake (node 1) and another one on a road (nodes 2).

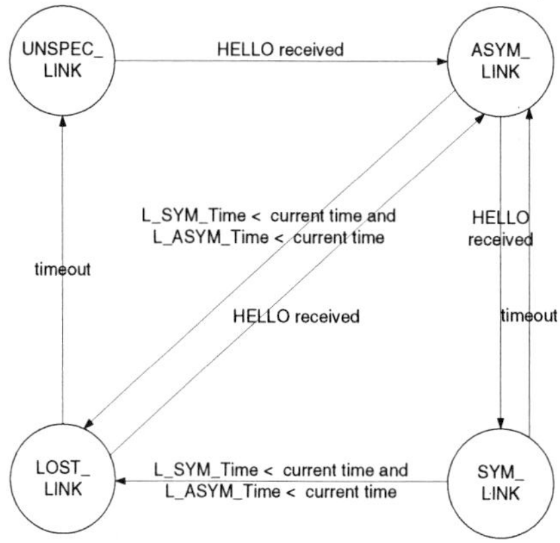

Figure 6.2.: Example for Discrete State Modeling: OLSR Link State

Examples of information that are modeled as *discrete states* are available resources or routing information. This includes, for examples, team members, vehicles, other equipment (e. g., water pumps), relief supplies and victims. In addition to the quantity of entities also specific characteristics may be relevant such as the particular skills and expertise of team members or the type of injuries of victims.

A network related example of discrete state information is shown in figure 6.2. The state of network links between network nodes, e.g., in a routing protocol such as OLSR [CJ03], can be modeled using a finite state machine. The current state can be described using a discrete value, e.g., asymmetric link (`ASYM_LINK`), and a transition may lead to another state within a limited set of discrete target states, e.g., lost link (`LOST_LINK`) or symmetric link (`SYM_LINK`).

All components of the MANET should be modeled and represented by a state model. Relevant (discrete and continuous) parameters need to be selected and used to model system states, e.g., routing protocol data, node positioning, and radio characteristics.

6.1.2. Probabilistic State Estimation

In this section we discuss probabilistic state estimation and the specific requirements in tactical MANETs. We present a particle filter based approach that can also handle error prone or false input data increasing the robustness of situation assessment in the targeted scenarios.

6.1.2.1. State Estimation in Tactical MANETs

The objective of the state estimation process is to derive an estimate of the current situation represented by system state values. This is an online process that should deliver approximations in real-time to the user. The estimator takes all currently available measurements as input data and calculates an approximation of the system state parameters. In the following we briefly outline the mathematical background which are required to present our probabilistic state estimation concept, more details and related work can be found in section 3.4.

Assumptions The target state evolves according to a discrete-time stochastic model defined by the following equation:

$$\mathbf{x}_k = \mathbf{f}_{k-1}(\mathbf{x}_{k-1}, \mathbf{v}_{k-1}) \tag{6.1}$$

where $\mathbf{f}_{k-1} : \mathbb{R}^{nx} \times \mathbb{R}^{nv} \to \mathbb{R}^{nx}$ is a known, possibly nonlinear function of the state $\mathbf{x}_{k-1}$ and $\mathbf{v}_{k-1}$ and $\mathbf{v}_k \in \mathbb{R}^{nv}$ a process noise sequence.

Objective The overall objective of our architecture is to recursively estimate $\mathbf{x}_k$ based on measurements $\mathbf{z}_k \in \mathbb{R}^{nx}$ that are related to the target state via the measurement equation:

$$\mathbf{z}_k = \mathbf{h}_k(\mathbf{x}_k, \mathbf{w}_k) \tag{6.2}$$

where $\mathbf{h}_k : \mathbb{R}^{nx} \times \mathbb{R}^{nw} \to \mathbb{R}^{nz}$ is a known, possibly nonlinear function of the state $\mathbf{x}_k$ and the measurement noise sequence $\mathbf{w}_k$. $\mathbf{w}_k \in \mathbb{R}^{nw}$ is the measurement noise sequence.

From a Bayesian perspective the objective is to recursively quantify some degree of belief in the state $\mathbf{x}_k$ at time k given data $\mathbf{z}_{1:k}$ up to time k. Thus it is required to construct the posterior PDF $p(\mathbf{x}_k|\mathbf{z}_{1:k})$ which can be obtained recursively in two stages: prediction and update.

6.1.2.2. Requirements

The following state estimation capabilities are required for robust situation awareness in tactical environments:

- *Comprehensive Applicability* The state estimation process should be generally applicable to all relevant application scenarios.
- *Incorporation of Constraints* It should be possible to incorporate any kind of information (e. g., mission information) that poses constraints on the likeliness of systems states in order to allow diminishing less likely or impossible states.
- *Discrete and Continuous States* The estimation mechanisms should be also applicable to hybrid system states including both discrete as well as continuious states.

The technological implications of each of these capabilities are described in more detailed in the following.

Comprehensive Applicability The state estimation system should be open and applicable to a wide range of application scenarios. This includes new, somehow unclea r or not-so-well-known scenarios with very limited prior knowledge. As described in the previous sections continuous and discrete states should be supported. It has to be able to deal with nonlinear state propagation as well as Non-Gaussian process noise. It should be able to incorporate multiple (in-)direct observations referring to the same state and also support nonlinear relationships between observations and system states.

Modeling and Incorporation of Constraints Different kinds of information such as mission information, domain-specific knowledge and general background information may be used to improve state estimation in tactical MANETs. This information may pose some constraints on the likeliness of specific systems states or even render them impossible (cf. section 4.2).

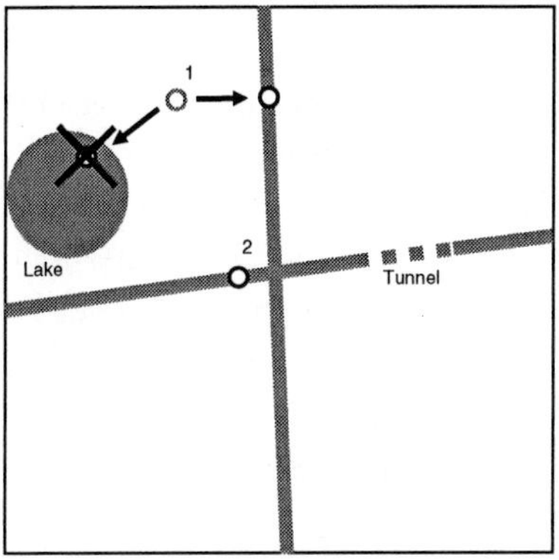

Figure 6.3.: Location Tracking Example: Incorporation of Constraints

Mission information may be utilized for these purposes, e. g., including operational planning, team composition, available equipment and (if applicable) vehicle types. Additionally, domain-specific background knowledge about standard procedures (e. g., typical tactical mobility patterns) and equipment categories and specifications (e. g., radio transmission characteristics, batteries) may be exploited. Furthermore general knowledge sources (e. g., topography model of operation environment) may pose some additional constraints on the state estimation process.

Constraints may either be strict constraints and basically render specific states impossible, e. g., impossible physical location due to physical laws, or may provide indication about the likelihood of specific system states, e. g., based on typical tactical mobility patterns. This kind of knowledge should be modeled as probability distributions to allow the inclusion of this type of information in a consistent way. Some of these constraints may be based on static and some on dynamic data sources.

Figure 6.3 shows a follow up of the location tracking example presented above (see a discussion of related work in the area of GMTI in section 3.4.3.4 on page 111). Two potential location tracks were followed in parallel for node 1, which represents a car. The track towards the lake leads to a location that is impossible (or at least very unlikely) whereas the other track leads to the road. This may lead to the conclusion that only the track towards the road is reasonable and the other track may be discarded.

6.1.3. State Estimation based on Particle Filters

Many probabilistic state estimation approaches have been proposed in the literature, e. g., Kalman filters. They are optimized for specific objectives and efficient, effective and most suitable in the respective application environments. *Particle filters* are the most suitable state estimation approach for robust situation awareness in tactical MANETs as they provide the best fit to the required capabilities described above.

We will discuss this in more detail in the following and present the specific characteristics of particle filters showing how they provide the required capabilities described above. Then we present the mathematical modeling that is used with in this section.

6.1.3.1. Characteristics

These are the specific characteristics of particle filters and the relevant capabilities:

- Particles filters are not limited to application scenarios with linear interrelationships. They support nonlinear state propagation functions and also non-Gaussian noise. Whole unknown a-posteriori probability density functions may be modeled and estimated using particle filters. Therefore they provide *comprehensive applicability* for state estimation in tactical MANET scenarios as requested in the previous section.
- Additional information may be elegantly integrated into a particle filter based state estimation processes, constraints can be incorporated at multiple stages. For example, the prior probability distribution used for the generation of new sample may be adjusted based on such information sources. The system update process may incorporate mission information, e. g., typical mobility patterns, and particles representing impossible states may be removed based on additional knowledge sources (see more details in section 3.4.3.4 on page 111). Overall particle filters provide flexible means for the *incorporation of constraints*.
- Particle filters are well suited for *hybrid state vectors comprising both discrete and continuous variables*. If the system includes discrete state variables the system models typically become nonlinear and non-Gaussian. Therefore, particle filters are suitable and have been widely used for hybrid state estimation [Kaw12].

Other probabilistic approaches such as Kalman filters may be feasible for specific subproblems and application scenarios but they do not provide the capabilities and flexibility discussed in the previous section. Therefore only *particle filters* are considered in the following and further elaborated within this dissertation. Other approaches may have some advantages such as higher performance for restricted application scenarios but none of them provides the overall required capabilities.

6.1.3.2. Mathematical Modeling

In the following we briefly review the mathematical background of state estimation using particle filters [DFG01, RAG04, EW09, EKW12]. A more detailed description of the particle filter concept (also known as sequential Monte Carlo method, SMC) can be found in section 3.4.2 on page 96.

The key concept of particle filters is to represent probability density functions by a set of samples (also referred to as particles) and their associated weights. The PDF $p(\mathbf{x}_k|\mathbf{z}_{1:k})$ is therefore approximated with an empirical density function (see also equation 3.32 on page 98):

$$p(\mathbf{x}_k|\mathbf{z}_{1:k}) \approx \sum_{i=1}^{N} \tilde{q}_k^i \delta(\mathbf{x}_k - \mathbf{x}_k^i), \quad \sum_{i=1}^{N} \tilde{q}_k^i = 1, \quad \tilde{q}_k^i \geq 0, \forall i. \tag{6.3}$$

where δ is the Dirac delta function and $\tilde{q}_k^i$ denotes the weight associated with particle $\mathbf{x}_k^i$.

Monte Carlo Integration Monte Carlo (MC) integration is the basis for SMC methods. Its objective is the numerical evaluation of a multidimensional integral [RAG04] $I(\mathbf{g}) = \int \mathbf{g}(\mathbf{x})\, d\mathbf{x}$ where $\mathbf{x} \in \mathbb{R}^{n_x}$. Monte Carlo methods factorize $\mathbf{g}(\mathbf{x}) = \mathbf{f}(\mathbf{x}) \cdot \pi(\mathbf{x})$ in such a way that $\pi(\mathbf{x})$ is interpreted as a probability density satisfying $\pi(\mathbf{x}) \geq 0$ and $\int \pi(\mathbf{x})\, d\mathbf{x} = 1$.

The assumptions is that it is possible to draw $N \gg 1$ samples $\{\mathbf{x}^\mathbf{i}, i = 1, \ldots, N\}$ distributed according to $\pi(\mathbf{x})$. An approximation of the integral $I(\mathbf{g}) = \int \mathbf{f}(\mathbf{x})\pi(\mathbf{x})\, d\mathbf{x}$ is the sample mean (see also equation 3.35 on page 99):

$$I_N(\mathbf{g}) = \frac{1}{N}\sum_{i=1}^{N}\mathbf{f}(\mathbf{x^i}). \tag{6.4}$$

Importance Sampling Ideally we want to generate samples directly from $\pi(\mathbf{x})$ and estimate I using equation (6.4). Suppose we can only generate samples from a density $q(\mathbf{x})$ which is similar to $\pi(\mathbf{x})$, then a correct weighting of samples makes SMC estimation still possible. The PDF $q(\mathbf{x})$ is referred to as the *importance* or *proposal density*, where $q(\mathbf{x})$ and $\pi(\mathbf{x})$ have the same support.

A Monte Carlo estimate of I can be computed by generating $N \gg 1$ independent samples $\{\mathbf{x}, i = 1,\ldots,N\}$ distributed according to $q(\mathbf{x})$ and forming the weighted sum (see also equation 3.38 on page 100):

$$I_N(\mathbf{g}) = \frac{1}{N}\sum_{i=1}^{N}\mathbf{f}(\mathbf{x})\tilde{q}(\mathbf{x}), \qquad \text{where} \quad \tilde{q}(\mathbf{x}) = \frac{\pi(\mathbf{x})}{q(\mathbf{x})} \tag{6.5}$$

are the importance weights.

Degeneracy Problem For an importance function as described above, it has been shown that the variance of importance weights can only increase over time. This may lead to the degeneracy phenomenon, i. e., many particles have negligible normalized weights. A suitable measure of degeneracy of an algorithm is the effective sample size N_{eff} (see also equation 3.45 on page 102):

$$\hat{N}_{eff} = \frac{1}{\sum_{i=1}^{N}(q_k^i)^2}. \tag{6.6}$$

$1 \leq N_{eff} \leq N$ where $N_{eff} = N$ if the weights are uniform and $N_{eff} = 1$ if one weight is $q_k^j = 1$ and all other are 0 (maximum degeneracy).

Resampling Resampling is required whenever a significant degeneracy is observed, i. e., when $\hat{N}_{eff}$ falls below some threshold $\hat{N}_{thr}$. Resampling eliminates samples with low importance and multiplies samples with high importance weights. It involves a mapping of random measure $\{\mathbf{x}_k^i, q_k^i\}$ into a random measure $\{\mathbf{x}_k^{i*}, 1/N\}$ with uniform weights. The new set of random samples $\{\mathbf{x}_k^{i*}, 1/N\}_{i=1}^N$ is generated by resampling (with replacement) N times from an approximate discrete representation of $p(\mathbf{x}_k|\mathbf{z}_{1:k})$ given by

$$p(\mathbf{x}_k|\mathbf{z}_{1:k}) \approx \sum_{i=1}^{N} q_k^i \delta(\mathbf{x}_k - \mathbf{x}_k^i) \tag{6.7}$$

so that $P(\mathbf{x}_k^{i*} = \mathbf{x}_k^j) = q_k^j$ (see also equation 3.46 on page 102).

As we want to apply particle filters for distributed and cooperative state estimation it has to be considered is that measurements can only be combined if they refer to the same state, and therefore the same instance in time. Only simultaneously measured observations can be directly combined. In the following section we analyze this problem in more detailed and introduce a distributed particle filter based approach that overcomes these issues using re-simulation whenever measurements need to be combined that refer to different moments in time.

6.2. Distributed Particle Filter based Approach

In this section we introduce our distributed state estimation approach based on particle filters [EW09]. We give and overview of the distributed state estimation architecture and describe the three main components in more detail: data selection and exchange, system update and situation assessment.

6.2.1. Distributed State Estimation Architecture

In the following we present our architecture for distributed state estimation that allows parallel computation of probability distributions based on decentralized storage of relevant measurement results and multisensor data fusion.

Nodes in a neighborhood cooperate in a distributed, concurrent computation process to get an improved estimate of the overall system state. This approach is more efficient and suitable for our scenario than central data pooling and fusion since it is less expensive in terms of a available communication bandwidth and energy.

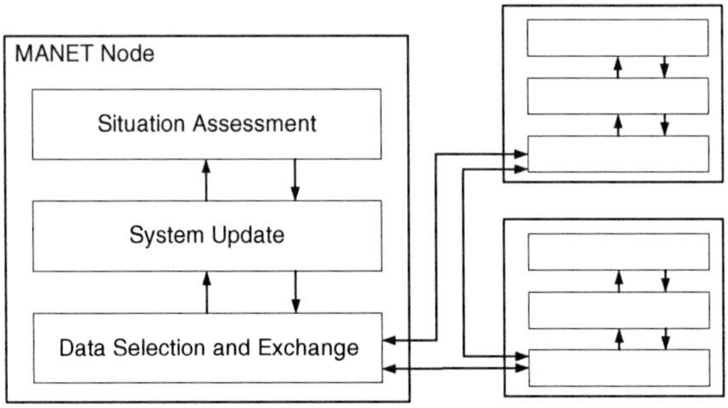

Figure 6.4.: Data Fusion Concept

The core components of our proposed architecture are shown in figure 6.4:

- *Data Selection and Exchange:* In this component data sets (representing observations and partial results) are determined which are most useful for other nodes within the distributed state estimation approach. The most relevant data are dynamically selected based on the current situation estimation and priorities derived from mission objectives according to specific metrics. The determined data sets are exchanged with selected other network nodes based on an efficient communication protocol.
- *System Update:* The current system state estimations represented by particle filters are updated based on all (newly) available information. This includes locally observations, data received from other network nodes as well as any additional information sources that provide constraints on the state estimation process. Resimulation is applied for the incorporation of measurements provided by other network nodes.
- *Situation Assessment:* The result is an estimation of potential system states which are analyzed, correlated and integrated in order to derive a comprehensive assessment of the current situation. The particle filter based approach provides probabilistic mechanisms which more robust in respect to inaccurate and error-prone data.

Each of these components is described in more details subsequently in the following sections.

In a distributed state estimation system the time that a specific measurement or state estimate refers to needs to be clearly defined. As tactical MANETs are a distributed and very dynamic system we do not expect that the system clocks of individual network nodes are precisely synchronized.

However, we assume a temporal ordering between individual measurements, e. g., based on Lamport clocks (see a more detailed discussion in section 6.2.3.2 on page 200). Indices always refer to the time of actual measurement and not to the time of arrival at an analyzing node in the following.

6.2.2. Data Selection and Exchange

Our goal is an efficient and resource-conserving communication protocol based on selective exchange of sensor measurements and partial results while providing the most relevant data for an effective state estimation. We therefore select the most suitable data that are exchanged with direct neighbor nodes to improve the local state estimation process.

Data Subset We assume that the node of interest has index $j = 1$, hence $\mathbf{z}^1_{1:k}$ is the local data acquired by node $j = 1$, whereas $\mathbf{z}^j_{1:k}$ for $j \neq 1$ refers to observations performed by other nodes. The concept is to use a subset $\zeta_{1:k}$ of all available information $\mathbf{z}_{1:k}$ to approximate the posterior PDF (cf. [RGT03]). Each set $\zeta^j_{1:k}$ $(j \neq i)$ is a subset of the total set of measurements acquired by node j:

$$\zeta^j_{1:k} \quad \subseteq \quad \mathbf{z}^j_{1:k} \tag{6.8}$$

whereas for $j = i$ the full set of local observations is assumed to be available ($\zeta^i_{1:k} = \mathbf{z}^i_{1:k}$).

The posterior PDF is approximated using these subsets of measurements as shown in the following for node $j = 1$:

$$p(\mathbf{x}_{1:k}|\mathbf{z}^1_{1:k}, \ldots, \mathbf{z}^N_{1:k}) \quad \approx \quad p(\mathbf{x}_{1:k}|\mathbf{z}^1_{1:k}, \zeta^2_{1:k}, \ldots, \zeta^N_{1:k}) \tag{6.9}$$

As it is assumed that the available data is partially redundant, Equation (6.9) should deliver a good approximation of the posterior PDF.

Information Content The most critical challenge is the selection of a suitable subset $\zeta^j_{1:k}$ for the data exchange. A candidate criteria for the selection process could be the information content, i. e., relative entropy and Kullback-Leibler divergence in order to select the most interesting and relevant data in the sense that it delivers most (new) information to the neighboring nodes (cf. [RGT03]).

Let $\mathbf{z}^j_\tau$ be a measurement in node j's local database and write $\mathbf{D}^1 = (\mathbf{z}^1_{1:k}, \zeta^2_{1:k}, \ldots, \zeta^N_{1:k})$ for the data available to node 1. Then the information of $\mathbf{z}^j_\tau$ relative to the 1st node's local belief is given by

$$q(\mathbf{z}^j_\tau) \quad = \quad D_{KL}\left(p(\mathbf{x}_{1:k}|D^1)||p(\mathbf{x}_{1:k}|D^1, \mathbf{z}^j_\tau)\right), \tag{6.10}$$

where D_{KL} denotes the Kullback-Leibler divergence. The maximum of this expression and therefore the most information content would be obtained for $\bar{z}^j_\tau = \mathrm{argmax}_\tau q(\mathbf{z}^j_\tau)$.

Data selection based on information content leads to a preference of information that is different. Therefore, malicious information provided by attackers may be preferred as falsified data may differ significantly from currently available state information. Data provided to neighbor nodes should in our architecture therefore also contain observations and information that confirms existing estimations.

Preferable Selection Criteria A more suitable selection criterion for our purposes is observations reliability and importance level for target nodes. Within MANET environments information provided by direct neighbor nodes may be considered particularly trustworthy and it is far more difficult to draw reliable conclusions regarding distant nodes. The selection algorithm should therefore prefer direct observations and information that refer to the direct neighborhood of the target node and gives less importance to information provided by distant nodes and observations that refer to distant nodes.

The data selection process should dynamically incorporate the current situation. Mission objectives as well as current state estimations should be used as a basis for the selection criteria. Data requests may focus on information regarding nodes of the direct neighborhood that are already known to update, i. e., confirm or alter the current state estimation regarding those nodes, or on aspects and topics about which not much data is available yet.

Data sets are dynamically selected and requested according the selection criteria described above and regularly exchanged with direct neighbor nodes. Incoming information is passed to the System Update module for further processing which is described in the following section.

The cooperation process is implemented in such a way that it does not introduce additional communication protocols which would significantly increase the network load. Instead they are conducted by extending the used routing protocol by piggybacking them on periodically exchanged broadcast messages, e. g., HELLO messages in Optimized Link State Routing (OLSR).

6.2.3. System Update

In this component the system is updated which comprises state estimation update (including incorporation of constraints) and re-simulation as described in the following two sections.

6.2.3.1. State Estimation Update and Incorporation of Constraints

The system state is updated based on all newly available measurements and partial results. The state may include a set of discrete state variables, e. g., characterizing a mode of operation or an object type in tracking systems. The particle filter based approach facilitates such hybrid state estimation scenarios.

The state estimation update also considers additional information sources for improved state estimation in tactical MANETs. These additional information sources include mission information, domain-specific knowledge and general background information.

State Estimation Update based on new Measurements Newly newly available measurements z_l are incorporated by updating the importance weights of the set of samples representing the current state x_{l-1} (*Importance Sampling*). The update calculation of the new weights q_i for the current samples $\{\mathbf{x}_l^i, i = 1, \ldots, N\}$ considers also additional information source which need to be reflected within the respective probability distribution. Additional information sources include mission information, e. g., typical mobility patterns, and other background information, e. g., topography model (cf. [RAG04, chapter 10 “Terrain-Aided Tracking”]).

Incorporation of Constraints Additional knowledge sources may pose constraints on potential system states. Strict constraints may render specific states impossible, e. g., impossible physical location due to physical laws. Samples $\mathbf{x}_l^i$ representing impossible states can be removed in order to improve the overall state estimation process.

In location tracking impossible states may be locations that are too far away from the newly measured location z_l, invalid locations according to a topography model or locations that are too far away from the center of a group formation (based on tactical mobility model constraints).

Particle Filter Consolidation As described in section 6.1.3.2 the variance of importance weights increases over time and after a number of steps only a few particles will have significant weight. Additionally, the number of particles may have decreased due to the incorporation of constraints.

Therefore the particle filter needs to be consolidated, i. e., the set of existing particles nees to be resampled and additional particles need to be generated.

6.2.3.2. Temporal Sequence of Measurements

Another important aspect of distributed data fusion is their temporal and causal relationship: How can measurements be combined and integrated that are received from different network nodes in particular if they were performed at different times?

Temporal and Causal Relationship of Data Sets: The semantic of events have to be considered to arrange measurements of different nodes in a distributed system (see section 3.4.3.3 on page 108 for a discussion of related work about out-of- sequence measurements (OOSM) in GMTI). A mechanism is needed that defines and identifies interdependencies between measurements based on a (partial) ordering that represents their temporal and causal relationship.

For these purposes logical clocks such as Lamport timestamps or vector clocks are used. Lamport timestamps [Lam78] provide a numerical capturing of a partial ordering (happened-before ordering) of events. Monotonically increasing counters C are maintained by each entity and are updated for each internal event and whenever a message is sent or received. This ensures a partial causal ordering of two measurements a and b of a node: if $a \rightarrow b$ then $C(a) < C(b)$, where $\rightarrow$ means "happened before". In addition vector clocks allow a total ordering of events in a distributed system. They work in a similar way as Lamport timestamps, but instead of a single counter a vector clock keeps an array of logical clocks, one for each entity. Every node keeps a local copy of this vector. The local clock is updated when a node receives information regarding an internal event or when it sends a message. When a node receives message the local clock is incremented and additionally each element of its vector clock is updated by taking the maximum of the value in its own vector clock and the value in the received message.

Re-Simulation As observations retrieved from other nodes may refer to a moment in time that is several steps in the past particles will have overlapping histories at a specific time t_k. Each node therefore keeps additionally to the current measurement values a history of received and local measurements for a specific period of time $\Delta t_{history}$.

When data arrives that refers to an earlier time τ (cf. [RGT03]) the respective particle filter is re-simulated. This means that a part of the filter's history is erased and it is run forward again to the current time t_k incorporating all collected observations at appropriate intervals. Measurements outside of the history window ($t_k - t_l > \Delta t_{history}$) are discarded. This leads to improved re-sampling of particles and improved performance of the particle filter. The particle filter needs to be re-simulated from time t_k whenever an observation is received for some time $\tau < t_k$.

6.2.4. Situation Assessment

The perception of various kinds of sensor data related to nodes and other resources within the tactical environment are the basis of the state estimation process. These observation may either be locally available or provided by other network nodes via the MANET.

The distributed and cooperative data fusion based on particle filters (as described above) utilizes this data and provides estimates of current systems states. Additional information (e. g. mission information) that poses constraints on the likeliness of systems states is incorporated into this process in order to allow diminishing less likely states. This provides some background on normal behavior for particular application scenarios. The aggregation and fusion of available state information enables the analysis of relationships between different entities and the comprehension of their meaning.

Using available sensor data and state information plausibility checks can help to detect deviations of individual measurements and identify suspicious nodes. Suspicious may be raised if an attacker changes routing information or maliciously modifies the routing algorithm in such a way that network parameters, e. g., distribution of sequence numbers or hop counts, differs from normal network operation.

Based on this analysis various working hypotheses are developed that explain the background of the current state estimation and measurements and describe the overall situation. This includes the projection of current state estimates describing the the (tactical) situation in the near future, e. g., development of the impact of a disaster or expected locations of task forces. Subsequently further analysis and investigation steps are triggered that may confirm or disprove a working hypothesis. The analysis process and fusion methods can be adapted to use the most suitable concept to identify deviant measurements or reports.

6.2.5. Distributed Particle Filter Protocol

The basic idea of the distributed particle filter concept is that each network node utilizes a set of particle filters to probabilistically model the state of the other network nodes. Each particle filter is initialized when first measurement arrive and continuously updated afterwards.

The following actions performed in each round, on each node for each of the other nodes:

1. Initialization (when first measurements arrive)
2. Prediction and Filtering

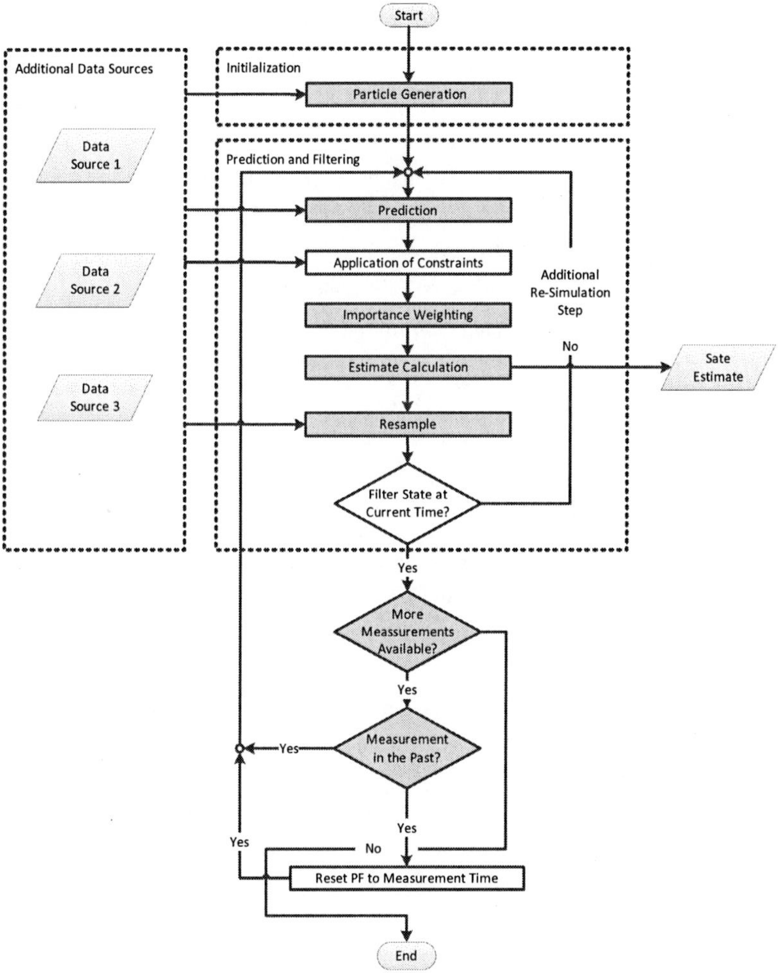

Figure 6.5.: Distributed Particel Filter Concept

An overview of these actions is shown in figure 6.5 and the individual steps of each action are described in more detail in the following sections.

6.2.5.1. Initialization

The particle filter for state estimation of another node is initialized at time t_k when the first measurements (referring to some time t_l) of that specific node arrive:

- *Generation of samples* $\{\mathbf{x}_l^i, i = 1, \dots, N\}$ equally distributed around newly available measurements $\mathbf{z}_l$, incorporating additional information sources.
- If required ($t_l < t_k$): *Re-simulation* from time t_l to current time t_k for this node based on system evolution model (cf. equation (6.1)), incorporating additional information sources.

6.2.5.2. Prediction and Filtering

All newly arriving incoming location messages are preliminarily processed. The output of this process is a set of new measurements z_l (referring to some time $t_l \leq t_k$ in the past). For some nodes i new measurements are available in a specific round and for other nodes not.

Based on the availability of new measurements z_l for another network node in a specific round the particle filter for a specific node is reset to a moment in time t_{l-l} in the past or not. Subsequently the system state is updated according to the procedure described below. The basic input for the prediction and filtering step is the set of samples $\{\mathbf{x}_{l-1}^i, i = 1, \dots, N\}$ representing the state of another node at a specific time $t_l < t_k$ (or if no new measurements are available l is equal to k).

- *System update* based on previous system state estimation x_{l-1} and system evolution model (cf. equation (6.1), incorporating additional information source (similar to re-simulation of one step).
- If new measurements are available:
    1. *Elimination of impossible system states*: Remove all samples that are invalid due to newly arrived measurement z_l^i.

2. If not enough samples are left: *Generation of new samples* $\{\mathbf{x}_l^i, i = 1, \ldots, N\}$ equally distributed around the newly available measurement $\mathbf{z}_l$, incorporating additional information sources.
3. If required ($t_l < t_k$): *Re-Simulation* from time t_l to current time t_k for this node based on system evolution model (cf. equation (6.1)), incorporating additional information source.

The goal of re-simulation is to generate new samples for the current time t_k $\{\mathbf{x}_k^i, i = 1, \ldots, N\}$ by probabilistically updating all previously calculated samples $\{\mathbf{x}_j^i, l \leq j \leq k, i = 1, \ldots, N\}$ stored in the history list up to estimation time t_k.

6.3. Probabilistic State Modeling for Task Force Tracking

In this section we present how our particle filter based approach can be applied to Task Force Tracking (TFT) as an example application for situation awareness in tactical environments. TFT s a tactical term for a localization system based on localization sensors, e. g., GPS, with the objective to provide location estimates of the location of task forces. The resulting system is our Sequential Monte Carlo method based TFT approach (SMC-TFT). It shows the major concepts and building blocks of probabilistic state estimation based on particle filters such as modeling of relevant states and additional information sources as well as MANET communication for distributed state estimation.

6.3.1. Task Force Tracking

TFT is an essential element to any tactical environment given its ability to contribute to situational awareness at all levels.

There are several challenges to TFT that render it a complex problem: Nodes (task forces) are typically scattered on the operation area, the environment changes dynamically due to node mobility and operations need to be coordinated not only within the task force of a specific organization, but also within other organizations. For rapidly evolving tactical situations in confined areas (such as those encountered in urban areas) existing mechanisms (e. g., satellite-based systems) are limited by several factors including update frequency, accuracy, and robustness to communication channel disruption (due to jamming or capacity limitations).

Satellite-based TFT systems highly depend on the availability of a complex infrastructure and backend. They are therefore vulnerable to attacks on the global infrastructure, e. g., jamming or denial of service (DoS) attacks. TFT systems based on mobile ad hoc networks (MANETs) that are locally deployed reduce the dependencies and can improve this situation.

MANETs in conjunction with sensors and information sources found in mobile computing components provide redundant, mutually supporting information on individual node locations. Data sources comprise geolocation (e. g., GPS) receivers, electronic compasses and gyroscopes, but also the ability to perform trilateration based on radio-frequency signals used for communication. The types of terrain and radio frequency environments to be expected in tactical MANETs implies that one cannot assume continuous connectivity across the MANET, but rather that the MANET will be partitioned frequently in an arbitrary manner.

Although MANET-based TFT systems have been proposed, these are typically limited to basic communication mechanisms that do not provide additional means to increase the robustness and the accuracy of TFT and are therefore inappropriate for challenging environments, e. g., urban operations or similarly constrained terrain.

6.3.2. Distributed TFT based on PF

The advantage of the proposed SMC-TFT mechanism is the combination of a particle filter based state estimation approach with MANET communication for distributed TFT. Robustness and accuracy of location estimation are increased within the proposed TFT system utilizing additional information sources such as mission information (group structure and tactical mobility patterns) and topographic data. This enables the combination of mobility models adapted to the tactical domain and the interpolation of likely node positions in the absence of immediate updates.

SMC-TFT is robust against the problems affecting naïve designs for MANETs or similar mesh-type networks described in the last section. It can be used as a substitute or supplement for existing infrastructure-based TFT systems.

In the following we describe some basic assumptions and present some examples of additional information that can be incorporated in order to enhance TFT in tactical MANETs. Then we show how re-simulation can be used to apply particle filters in distributed environments.

6.3.2.1. Basic Assumptions

The overall objective of our proposed SMC-TFT approach is to derive a probabilistic estimation of a node's location and velocity for enhanced TFT. Location and velocity are information examples that are more accurately modeled by continuous states. The system state vector could be complemented with a set of discrete state variables characterizing, for example, a mode of operation, a behavioral pattern or an object type. A hybrid state estimation approach could be used to model system dynamics in different discrete behavioral modes.

In order to keep the system clear and to show the main ideas and concepts of SMC-TFT we model the current state as two continuous state vectors: $\mathbf{x}_k^{loc}$ representing the position (x, y, z) and $\mathbf{x}_k^{vel}$ representing the velocity (v_x, v_y, v_z) of another node (cf. section 6.1.1). GPS sensors provide a mesurement of the geo-coordinate $\mathbf{z}_k^{loc}$ and the velocity $\mathbf{z}_k^{vel}$ of a node.

SMC-TFT State Evolution Model Location and velocity state variables evolve according to a discrete-time stochastic model where the system state variable $\mathbf{x}_k$ represents both node location $\mathbf{x}_k^{loc}$ and velocity $\mathbf{x}_k^{vel}$ (cf. equation 6.1).

GPS Measurement Equation The of sensor measurements $\mathbf{z}_k$ comprises the location measurement $\mathbf{z}_k^{loc}$ and the velocity measurement $\mathbf{z}_k^{vel}$ of the GPS sensor. The measurements noise sequences $\mathbf{w}_k \in \mathbb{R}^{n_w}$ are modeled as Gaussian white noise with mean zero and standard deviations σ_{GPS}^{loc} and σ_{GPS}^{vel} respectively (cf. equation (6.2).

6.3.2.2. Incorporation of Additional Knowledge Sources

An important tenet of the SMC-TFT concept is the inclusion of additional knowledge sources such as information about the tactical mission and a topographical model of the environment.

Mission Information Knowledge of mission objectives, participants and group structure of a specific mission are promising information sources for improving TFT. Typical formation patterns and typical (minimum, average and maximum) velocities can be derived from this form of information. Such properties may either belong to specific nodes or reflect a specific tactical situation. An example of how group structure and movement formation information can be modeled is the Reference Point Group Mobility Model (RPGM).

Topographical Model A topographical model of the environment can be used to improve estimation of node movement. The direction of movement is probabilistically dependent on properties of the area of movement [RAG04]. There are areas with restricted movement (e. g., water, mountains), areas of free movement (e. g., open field) and areas which are characterized with a high likelihood of following specific paths (e. g., roads, pathways, or stairs).

Additional information sources enable the preference of more likely state values and the exclusion of implausible estimations. For example, impossible locations can be removed from the set of potential locations based on node type and topography model, e. g., lakes or buildings for regular cars. Additionally impossible velocity values can be excluded based on node properties and physical laws, e. g., based on upper bounds for acceleration.

6.3.2.3. Re-Simulation

As decribed in section 6.2.3.2 on page 201 location observations retrieved from other nodes may refer to a moment t_l in the past. In order to enable re-simulation in this case a history of received GPS measurements is kept. Whenever such data arrives that is more recent than previously received measurements the respective particle filter is re-simulated. In this case the filter's history is erased and re-run to the current time t_k, incorporating all collected GPS measurements at appropriate intervals. Measurements that refer to a moment t_l in the past that are outside the history window ($t_k - t_l > \Delta t_{history}$) are discarded.

6.3.3. SMC-TFT Protocol

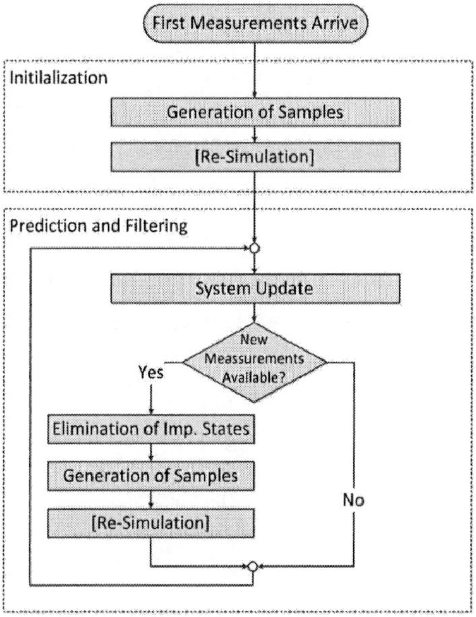

Figure 6.6.: Basic Concept of SMC-TFT

An overview of the SMC-TFT protocol is shown in figure 6.6. The basic concept is the distributed particle filter protocol (as described in 6.2.5 on page 202):

1. Initialization (when first measurements arrive)
   - *Generation of samples* incorporating additional information sources:

    - topography model for deletion of invalid samples or alignment of samples,
    - mission information for group movement patterns.
  - If required: *Re-simulation*
    Probabilistically updating all previously calculated samples stored in the history list up to estimation time, e. g., applying the mobility model and incorporating mission information and topographical model.
2. Prediction and Filtering
  - *System update*
  - If new measurements are available:
    a) *Elimination of impossible system states*: Remove all samples that are invalid due to newly arrived measurement z_l^i, e. g., that are
      - too far away from the measured location,
      - in an invalid location according to a topography model or
      - too far away from the center of a group formation.

    b) If not enough samples are left: *Generation of new samples*

    c) If required: *Re-Simulation*

The SMC-TFT approach was implemented in a MANET simulator based on [HE04] and evaluated. Simulation results show that the proposed SMC-TFT model outperforms two reference TFT models in both accuracy and robustness. More details and evaluation results are presented in the next chapter (see section 7.4.2 on page 258).

6.4. Summary

In this chapter we presented a concept for *probabilistic state modeling and estimation* [EW09] for situation awareness in tactical MANETs to increase the overall robustness regarding inaccurate and error-prone input data. The probabilistic modeling of system states provides increased flexibility and robustness as estimations are not restricted to a specific value but instead are represented by a “probability cloud”.

The proposed *concept based on Particle Filters is open and applicable to a wide range of application scenarios* (including nonlinear state propagation as well as Non-Gaussian process noise). It can also be applied for hybrid state estimation where the system state comprises both discrete and continuous variables. In this case discrete variables may characterize a mode of operation enabling system dynamics modeling based on different discrete behavioral modes.

The proposed state estimation concept allows the *incorporation of multiple observations and additional information sources* [EKW12], e. g., mission information, domain-specific background knowledge and general information sources. The particle filter based state estimation process enables the incorporation of additional information at multiple stages. Additional information sources may pose some constraints on the state estimation process and are exploited to adjust the likelihood of specific system states or exclude impossible states.

We presented a distributed state estimation architecture [EW09] which addresses the challenges resulting from the distributed nature of tactical MANETs. Observations and state estimates are exchange within the MANET but information received from other network nodes may refer to a moment in time in the past. Therefore we propose a *re-simulation mechanisms* for resolving temporal and causal correlation of observations and estimations performed by other network nodes.

The proposed concepts are applied to *Task Force Tracking* [EKW12], a typical application for tactical situation awareness. The implementation contains the major building block described above serves as a proof of concept for our probabilistic state modeling approach.

Part III.

Evaluation of Research Results

7. Simulation and Evaluation

In this chapter we evaluate the concepts that were presented in the previous chapters. For the implementation and validation we select suitable tools that allow showing the most important aspects in an efficient and effective way. In the following the simulation environments that were chosen for the evaluation and further develop are presented. Then the evaluation scenarios for all three main contribution area (cross data analysis, cooperative trust assessment and probabilistic state modeling) are presented and the evaluation results are discussed in detail. Finally the evaluation results are summarized and some conclusions are drawn.

7.1. Simulation Environments

In the following we describe the selection of suitable tools for the evaluation of our proposed solutions in the three research contribution areas: cross data analysis, cooperative trust assessment, and probabilistic state modeling. The objective is to show the most important aspects based on field trials and simulations in order to compare the results with related work and theoretical assumptions. Simulation environments should provide the required features or modular extensibility in order to efficiently and effectively simulate realistic application scenarios.

The different areas require different tools and simulation environments in order to simulate and evaluate the relevant properties:

- For the evaluation of the proof of concept for location related data sources of the *cross data analysis* concept the respective MANET properties and mechanisms need to be simulated, including node mobility, attacker models, wireless communication, and the incorporation of a topography model.

- For evaluation of the *cooperative trust assessment* we need mechanisms for (piggy-back) exchange of trust information, trust combination and forwarding.
- For the evaluation of the task force tracking proof of concept of the *probabilistic state modeling* we additionally need the means for particle filter simulation.

Overall the simulation environments need to provide the following capabilities and features in order to allow for realistic simulation of tactical MANETs and state estimation.

- *Wireless ad hoc communication mechanism* including common MANET routing protocols and radio propagation models (cf. section 3.2.4.2 on page 68).
- *Node mobility mechanisms* including random movement, group mobility models, and incorporation of topography model (cf. section 3.2.4.1 on page 64).
- *Modular extensible data processing and communication mechanisms* allowing local data processing and (piggy-back) exchange of additional data, e. g., trust data.
- *Attack mechanisms* including active attacker models, e.ġ., black hole and wormhole attacks, to show robustness and attack detection capabilities of proposed concepts (cf. [EB06, BETJ06, EP09]).
- *Particle filtering mechanisms* in order to evaluate the probabilistic sate modeling concept.

In general, network simulation is a very important tool for the development and validation of new methodologies in MANETs as it is hard to set-up real testbeds of a reasonable size incorporating node mobility and realistic radio propagation. Therefore, a variety of simulation environments has been implemented and is available for the research community. In order to cover all required features and capabilities (described above) for the evaluation of the proposed concepts we selected three simulation tools. In the following the selected simulation environments are described in more details.

- *ns-2* [GV] is a network simulator that provides various implementation of MANET routing protocols, mobility and radio propagation models. ns-2 is available as Open Source and is used by many research groups around the world. It is based on the programming languages C++ and OTCL, an object-oriented scripting language. ns-2 is a discrete network simulator where packets are the smallest units considered. The discrete packet modeling offers an event-driven perspective. ns-2 is based on the TCP/IP communication model and implementations of common MANET routing protocols (such as AODV and OLSR) are available. ns-2 includes an implementation of several radio propagation models and enables the modular extension with a topography based radio propagation model.

- *JiST/MobNet* [KBHS07] is a Java based network simulator based on JiST/SWANS, which was developed by Rimon Barr [Bar04a] and is available as Open Source. One of the main advantages of JiST/MobNet is that it is much more efficient than other popular simulation environments and can therefore be used to simulate big networks. It provides a number of routing protocol implementations and offers flexible and powerful configuration tools. JiST/MobNet enables the modular extension with data processing modules and (piggy-back) exchange of additional data, modular. Additionally, it provides an implementation of several attacker models, e. g., black hole and wormhole attack.
- The *PF Network Simulator* is based on a simulator that was developed and used in [HE04] (also used and extended in [BL08]). It is implemented in Java and enables network and localization simulations using Particle Filters. It provides the additionally required particle filtering mechanisms required for the evaluation of the probabilistic state estimation concepts.

In the next sections these network simulation environments are presented in more detail.

7.1.1. ns-2

The network simulator ns-2[1] [GV] is a comprehensive and freely available simulation environment. ns-2 includes an implementation of several radio propagation models, e. g., free space, two-ray ground and shadowing model (cf. section 3.2.4.2 on page 68) and enables the modular extension with a topography based radio propagation model. Additionally, there are movement generators, e. g., BonnMotion[2], for several mobility models, e. g., random waypoint model, Manhattan grid model and reference point group mobility model available (cf. section 3.2.4.1 on page 64).

[1] `http://www.isi.edu/nsnam/ns`
[2] `http://net.cs.uni-bonn.de/wg/cs/applications/bonnmotion/`

The development of ns-2 started already in 1989 with a focus on wired networks. ns-2 was later on extended with functionalities for WLAN and MANETs. ns-2 is open-source project and therefore everyone can validate the existing code, modify and expand it according to specific needs. iNSpect[3] [KCMC05] is a tool which can be used to visualize simulation results of ns-2. In the following we give an overview of the architecture and implementation of ns-2 and iNSpect.

7.1.1.1. Architecture and Overview

The design of ns-2 is based on packets which are the smallest units. The physical layer and the data link later are not modeled separately in ns-2, but form (as in the TCP/IP model) a common network access layer, which is used for the transmission of packets. The physical transmission of data is modeled as discrete events.

In the simulation environment nodes are aware of all packets in the network in contrast to real wireless networks and they decide if they are actually able to receive a specific packet or not. For these purposes suitable radio propagation models are required.

The implementation of ns-2 is based on the programming languages C++ and OTCL. C++ is a compiled object-oriented programming language which provides high efficiency for the calculation and processing of large amounts of data. OTCL is an object-oriented extension of the scripting language TCL and in contrast to C++ an interpreted language. OCTL allows easy and quick changes in the program code and is therefore used for the description of simulation scenarios within ns-2.

The results of a simulation that was performed using ns-2 is stored in a so-called *trace file*. A trace file is a plain text file that contains all relevant actions that were carried out during a simulation and additional information associated with the simulation timestamps. Trace files provide the basis for further evaluations of the simulation results.

[3] `http://toilers.mines.edu/Public/Code/Nsinspect.html`

7.1.1.2. Visualization Tool iNSpect

ns-2 does not include a graphical user interface. It is operated via a command-line interface and simulation results are written to trace files. However, a visualization tool can help to understand the content of trace files quicker and more easily and it allows an interactive analysis of simulation results.

ns-2 provides a visualization tool called nam[4] which is a Tcl/TK based animation tool for viewing network simulation traces. It supports topology layout, packet level animation, and various data inspection tools. However, it was originally developed for wired networks and is not suitable for the analysis of mobile networks.

iNSpect [KCMC05] is a visualization tool which was developed to visualize ns-2 simulation results of mobile network scenarios. It includes also the visualization of node movements and radio transmissions. iNSpect was written in C+ + using OpenGL.

7.1.2. JiST/MobNet

Additionally we use the simulation environment JiST/MobNet [KBHS07] which is based on the work of Rimon Barr [Bar04a] and was further extended at TU Darmstadt [KBHS07]. JiST/MobNet provides different fading and path loss models (cf. section 3.2.4.2 on page 68) as well as several mobility models, e. g., random waypoint model (cf. section 3.2.4.1 on page 64). We use JiST/MobNet for the evaluation of the developed concepts as it allows the modular extension with local data processing and (piggy-back) exchange of additional data.

Furthermore, JiST/MobNet provides several attack models, e. g., black hole and wormhole attack, that were implemented within the Diplom thesis supervised by me of Stefan Endler [End08] and evaluated in the Bachelor thesis supervised by me of Malcolm Parsons [Par08, PE09, EP09].

7.1.2.1. JiST

Java in Simulation Time (JIST) is a Java-based discrete event simulator and is much more efficient than other popular simulation environments. A more detailed descrip-

[4] `http://www.isi.edu/nsnam/nam/`

tion of the functionalities and performance results can be found in the JiST User Guide [Bar04c]. JIST by Rimon Barr as part of his PhD thesis [Bar04a]. Figure 7.1 [Bar04b, EHK*07] shows how Java source code is is converted by the javac compiler in Java bytecode. This bytecode is then converted by the *JIST rewriter* in modified bytecode, which is then processed by the simulation kernel. As a result of the Java based approach, JIST should generally run on all systems that provide a Java Virtual Machine (JVM), including Windows, Linux and Mac OS.

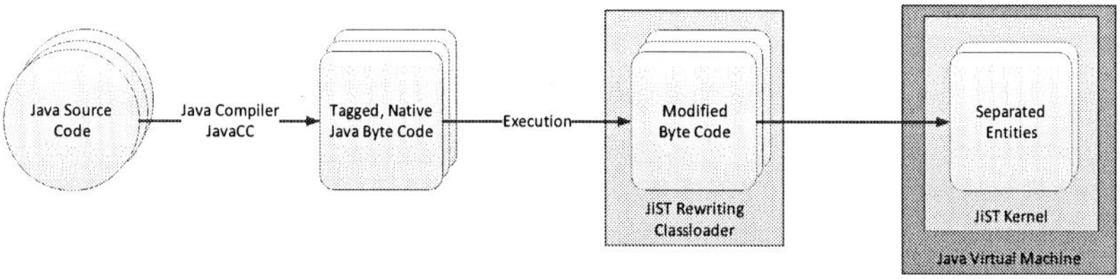

Figure 7.1.: Processing Chain of JiST (based on [Bar04c, EHKP06, EHK*07])

7.1.2.2. MobNet

The *Scalable wireless ad hoc network (SWANS)* which was also developed by Rimon Barr is a network simulator package that uses JIST to simulate MANETs. A more detailed description of SWANS can be found in the user manual [Bar04d]. SWANS biggest advantage, compared to other network simulators such as ns-2, is that it can simulated MANETs with a very large amount of nodes. SWANS provides an implementation of the routing protocols: AODV, Dynamic Source Routing (DSR) and Zone Routing Protocol (ZRP).

MobNet [EHK*07, KBHS07] is a further development of SWANS and also runs on the JIST kernel. The JIST kernel was extended to perform high performance simulations. The JIST/MobNet kernel supports real-time events, parallel event execution and a transparent sharing mechanism of simulation models on multiple computers.

The protocols that were available in SWANS have been validated and improved. In addition, some modified and extended versions of AODV and an implementation of the OLSR routing protocol are available. MobNet also provides, in addition to SWANS, an extended setup environment with significantly more options for simulations, traffic generation, online and off-line monitoring and analysis tools (see figure 7.2 [EHK*07]).

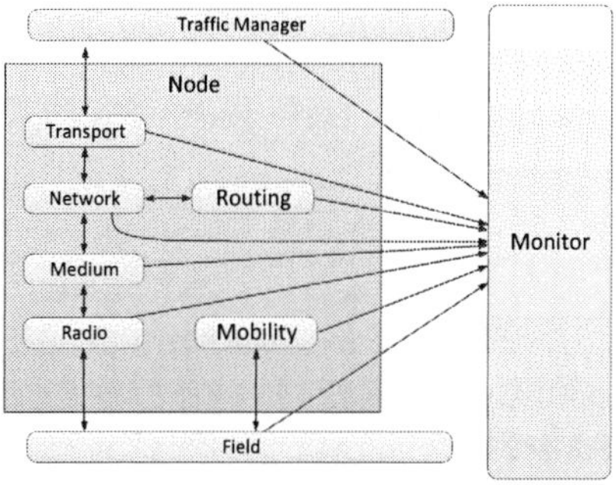

Figure 7.2.: Component-based Architecture of MobNet (based on [KBHS07,EHK*07])

7.1.3. PF Network Simulator

The *PF Network Simulator* is based on a simulator that was developed by Hu and Evans [HE04] and is available online via their Web site[5] at the University of Virginia. They used it to conduct experiments for validating their proposed localization protocols. The simulator is implemented in Java and supports seed nodes that know their location and other nodes (without localization sensors). Baggio and Langendoen reused and extended this simulator in [BL08].

Time is modeled as discrete steps in the simulator and the number of samples maintained may be varied. More samples may improve accuracy but require more memory and computation resources. For a given set of parameters, the simulator generates a number of random network configurations. This set of configurations is then executed and relevant simulation information, e. g., estimation errors, are calculated and logged for further analysis.

This simulator is used as a basis for the evaluation of PF based concepts and significantly extended within this work.

[5] `http://www.cs.virginia.edu/mcl`

Mobility Model The simulator provides two mobility models (cf. section 3.2.4.1 on page 64): *Random Waypoint Mobility Model* and *Reference Point Group Mobility model*. The basic idea of the Random Waypoint Mobility Model is that each node may randomly choose a destination, a movement speed and its pause time after arriving at the destination [CBD02]. The Reference Point Group Mobility model is used to investigate the effect of group behavior [HGPC99]. The motion of a node is modeled as the combination of a group motion vector and a random individual motion vector. It can be considered, for example, as an approximation for a group of nodes moving within a current.

For the evaluation we additionally implemented the *Manhattan Grid Model* [Spe98] based on the BonnMotion [AEGPS10] algorithm. The Manhattan Grid Model is based on road topology. Roads are located in a grid structure and nodes move in horizontal or vertical directions on these roads. The model follows a probabilistic approach: each node probabilistically chooses at each intersection to keep moving in the same direction or to turn left or right.

Radio Propagation Model The PF Network Simulator provides a basic deterministic radio propagation model based on transmission distance. This can be considered as an implementation of the Free Space Model where only line-of-sight radio transmissions through free space (without any obstacles, reflection or diffraction disturbances) are taken into account (cf. section 3.2.4.2 on page 68).

For the evaluation we additionally implemented the *Shadowing Model* based on ns-2 [FV11]. This model consists of two distinct parts: a path loss model (which predicts the mean received power) and a second that reflects the variation of the received power at a certain distance.

7.2. Cross Data Analysis: Location Related Information

In this section we present the experiments and evaluation results for the cross data analysis of location related information. First we evaluate the interrelationship of node distances and radio signal characteristics performing radio signal strength measurements in relation to distance in various environmental setups [AE08a]. Then we describe the implementation and evaluation of the topography based radio propagation model [REKW11] and present some additional experimental results for setups reflection and diffraction about obstacle. The last part concludes the section with the implementation and evaluation of the consistency checks for location based attack detection (see details in section 4.3.3 on page 141) [ES08].

7.2.1. Distance Estimation based on Radio Signal Characteristics

In this section some basic experiments for signal strength measurements in relation to distances are described. The goal is distance estimation based on radio signal characteristics under realistic conditions. Therefore a realistic test environment was chosen for the experiments instead of a sterile test environment, e. g., in a shielded laboratory, and common hardware to implement a typical and realistic scenario.

These experiments were executed within the Bachelor thesis supervised by me of Stefan Appel [App06] and the results were further elaborated and published in a joint paper [AE08a].

7.2.1.1. Experimental Setup

Measurements with different distances between two notebooks were performed in various environmental settings.

Equipment As receiver and transmitter two *Dell Latitude D600* notebooks with different WLAN cards (as will be described later) were used. The operating system was *Windows XP Professional SP2* and the wireless network options were set to ad hoc mode. For measuring and recording the signal strength *Netstumbler 0.4.0*[6] was used. This software can be used to record RSSI and associated SSID twice per second. It also supports GPS devices so that latitude and longitude can also be recorded.

Experiments In the following section signal strength measurements are presented. One notebook was used as stationary node and the second notebook as mobile node, equipped with GPS and *Netstumbler* to record the RSSI for different distances between

[6] `http://www.netstumbler.com`

both nodes. The distances were measured in two separate ways, one way using a measuring tape and another using GPS data and calculating the distance using spherical trigonometrical methods.

Measurements with different notebook orientations, surfaces and WLAN cards were taken. Table 7.1 gives an overview of each experiment. Table 7.2 shows general parameters for the experiments. Differences will be described separately for each experiment.

Table 7.1.: Experimental Setups

Experiment 1	
Description:	Different antenna orientations
Environment:	Undeveloped area
Line of sight:	√
Distance Measurement:	GPS
Ground:	Stubble field
Experiment 2	
Description:	Different surfaces
Environment:	Undeveloped area
Line of sight:	√
Distance Measurement:	GPS
Ground:	Stubble field
Experiment 3	
Description:	Different WLAN cards
Environment:	Undeveloped area
Line of sight:	√
Distance Measurement:	Measuring tape and GPS
Ground:	Stubble field

Table 7.2.: Equipment Used for the Experiments

Stationary Notebook	
WLAN card:	Cisco Aironet 350 (PCMCIA)
Altitude:	80cm
Measuring Notebook	
WLAN card:	D-Link DWL 660 (PCMCIA)
Altitude:	100cm
Orientation:	to each other

7.2.1.2. Results

In the following the results for the experiments specified above are presented.

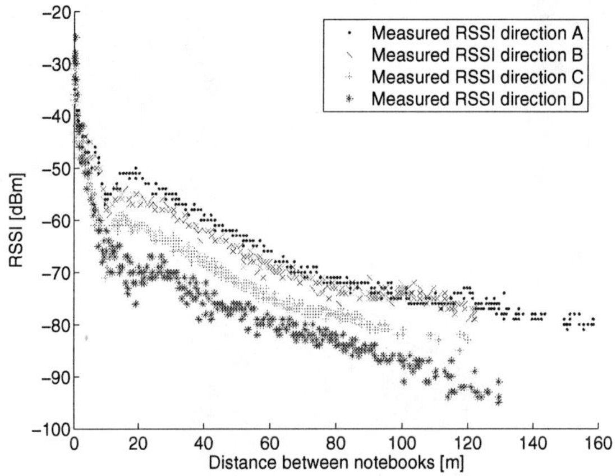

Figure 7.3.: Experiment 1: Different Antenna Orientations

Experiment 1: Antenna Orientation *Description* The distance between notebooks as well as notebook orientation influences measured RSSI. To analyze the influence of different horizontal orientations a further was performed. Figure 7.4 shows different orientations of both notebooks in relation to each other.

Evaluation As shown in figure 7.3 orientation of the notebooks is a significant factor for measured RSSI. The highest signal strength was measured when the notebooks were oriented perpendicular to each other (orientation A and B) while lowest signal strength was measured while the notebooks were oriented behind each other. These results show that antenna orientation and how it is integrated into the hardware has a significant impact on measurement results but the relationship seems to reflect a stable offset.

Experiment 2: Different Surfaces *Description* The surface on which the signal propagation takes place also has an effect on measured RSSI values. In order to collect data related to signal propagation on different surfaces, two measurements were conducted. One experiment was conducted on asphalt as a surface and another used a stubble field as surface.

(a) Orientation A: Perpendicular to Each Other (Orientation B: mirrored)

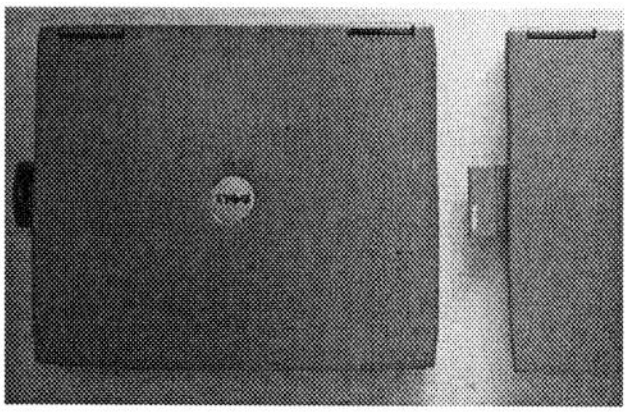

(b) Orientation C: Behind Each Other

(c) Orientation D: Towards Each Other

Figure 7.4.: Notebook Orientations in Relation to Each Other

Evaluation Figure 7.5 shows the correlation between distance and measured RSSIs on two different surfaces. It can be inferred from this figure that signal attenuation on asphalt is smaller than on a stubble field which leads to a flatter signal strength curve and results in a higher usable range. Furthermore both signal strength curves show a minimum at about 20 meters as a result of the two-ray propagation. These two measurements show the need of adapting a signal propagation model to environmental conditions. Besides the measured values, additional values calculated with the free space propagation model are plotted. To achieve the best results with the model the parameter $n = 2.5$ was chosen.
Another important input for the purpose of distance estimation based on signal strength measurements is highlighted in figure 7.5. Notice the high degree of scatter of measured RSSIs at high distances which makes it difficult to perform a distance estimation since one RSSI value may belong to a number of different distances leading to inaccurate predictions.

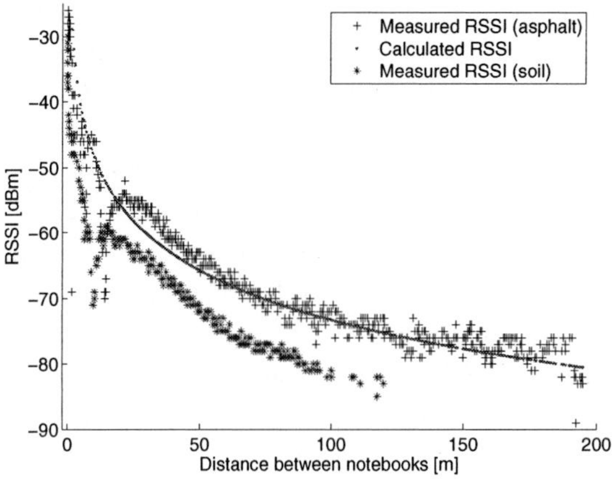

Figure 7.5.: Experiment 2: Different Surfaces

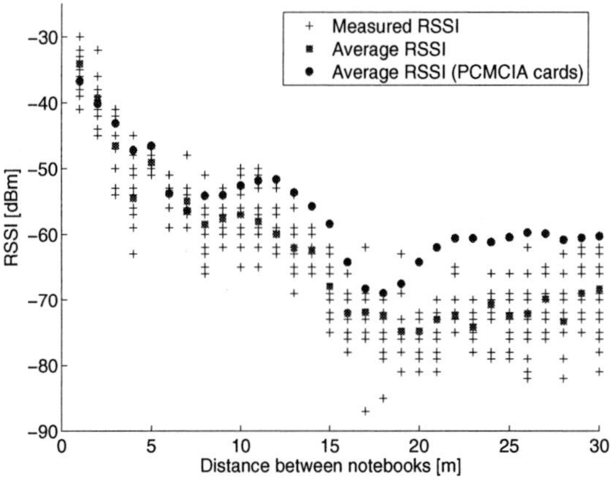

Figure 7.6.: Experiment 3: Different WLAN Cards

Experiment 3: Different WLAN Card *Description* The experiments conducted above used PCMCIA WLAN cards as stated in table 7.2. To investigate the impact of the use of a specific WLAN card model, an additional experiment was performed comparing effects on the radio signal with different hardware components. For this experiment the stationary notebook used an internal *Intel(R)-PRO/Wireless-LAN-3A* mini PCI card and the mobile measuring notebook used an internal *Dell Truemobile 1300* mini PCI card.

Evaluation The results of RSSI measurements using internal WLAN cards are shown in figure 7.6. Crosses show all recorded values while red squares are the average of these values. The circles in figure figure 7.6 represent RSSIs recorded using PCMCIA WLAN cards under identical environmental conditions. These two sets of values when compared against each other show the gradient of the curves are similar except for very small distances. At high distances RSSIs measured with internal cards are below the values measured with PCMCIA cards. This could be to the result of a number of factors such as antennas of internal cards could have different characteristics than external antennas or the RSSI value provided by the Intel card is not calibrated with the same settings as the D-Link card. This experiment shows that the type of WLAN card used has an important impact on the quantitative results of RSSI measurement, although qualitative characteristics remain the same. Therefore the propagation model must be adapted to the specific hardware used.

In general, the theoretical model and measured data reflect the same behavior: some oscillation with maxima and minima in the short range and a smooth decay in the far regions. However minima are not at the distances predicted by the two ray propagation model (at around 3, 7 and 13 meters), instead they appear at around 7 and 17 meters. This shows that the model is too simplistic for an exact prediction although it qualitatively shows expected physical effects. This is not surprising since the physical model is idealized and is based on some assumptions (such as an ideal metallic surface of the reflecting ground, which was not present during our experiments).

7.2.1.3. Using Signal Strength for Distance Estimation

Using outcomes from the previous signal strength measurements an application was developed to estimate the distance between two notebooks using measured RSSI values.

Theoretical Background To calculate a distance based on measured signal strength a propagation model with a injective mapping is necessary. Therefore logarithmic equation as expressed by the free-space propagation model was chosen.

Transforming the respective equation (cf. equation 3.1 on page 68 leads to the following simplified representation of the Free Space model based on a logarithmic scale:

$$P(d)[dBm] = P_{d_0}[dBm] - 10 \cdot n \cdot log\left(\frac{d}{d_0}\right) \quad , \tag{7.1}$$

where P_{d_0} in the known signal strength value for a specific distance d_0 from the sender and the parameter n describes the environment. $n = 2$ is a typical value for environments without obstacles.

This equation can be further transformed in order to directly calculate the distance:

$$d = 10^{\left(\frac{P(d_0)-P(d)}{10 \cdot n}\right)} \cdot d_0 \tag{7.2}$$

Equation 7.2 can be calibrated by performing two measurements in different but known distances between the two nodes. The first measurements determines $P(d_0)$ and d_0 and n can be determined using equation 7.3.

$$n = \frac{P(d_0) - P(d_1)}{10 \cdot \log\left(\frac{d_1}{d_0}\right)} \tag{7.3}$$

For the calibration and distance estimation a minimum distance should be considered which depends on the height of receiver and transmitter and can be calculated using the developed software.

Implementation The application was developed using the *Microsoft .NET 2.0 Framework* and is written in *Microsoft Visual Basic*. It uses the *Windows Management Instrumentation* (WMI) to read RSSI values from the WLAN card.

Figure 7.7 shows a screenshot of the application. The displayed distance is calculated based on RSSI values which were collected over a period of five seconds. These ten values are shown at the bottom of the application window and are updated every 0.5 seconds. The use of an average of ten different values reduces the influence of signal

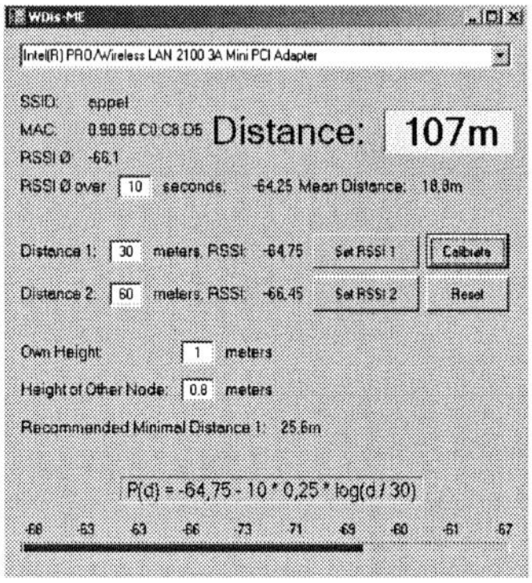

Figure 7.7.: Screenshot of the Application

fluctuations. As described above the application can also be calibrated and parameters in current use for the propagation model are shown.

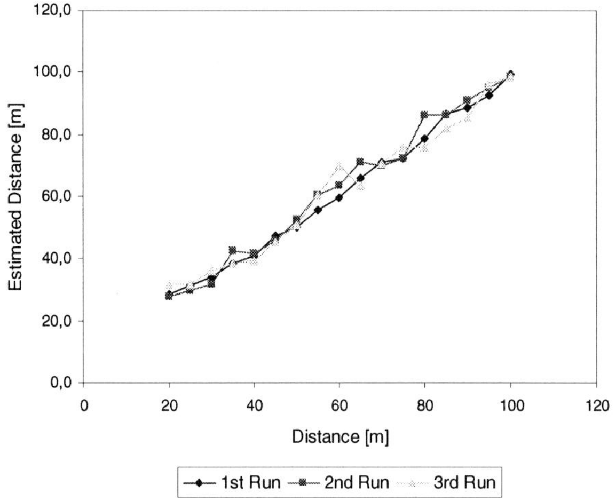

Figure 7.8.: Distances Estimated Using Application vs. Real Distances

Table 7.3.: Measurement Results of Application Evaluation

Distance [m]	**Estimated Distance [m]**			**Deviation [m]**	
	1st Run	2nd Run	3rd Run	Average	Maximum
20	28.7	27.9	31.9	9.5	11.9
25	31.5	30.0	31.9	6.1	6.9
30	34.3	31.9	36.0	4.1	6.0
35	38.6	42.5	38.4	4.8	7.5
40	41.0	41.6	39.5	1.0	1.6
45	47.3	45.7	45.1	1.0	2.3
50	50.1	52.5	50.8	1.1	2.5
55	55.8	60.4	60.4	3.9	5.4
60	59.6	63.7	69.9	4.7	9.9
65	65.9	71.3	63.7	2.8	6.3
70	71.2	69.9	70.8	0.7	1.2
75	72.5	72.2	75.7	2.0	2.8
80	78.8	86.4	75.7	4.0	6.4
85	86.7	86.4	81.9	2.1	3.1
90	88.8	91.1	85.3	2.3	4.7
95	92.6	94.8	96.1	1.2	2.4
100	99.4	98.7	98.7	1.1	1.3
Deviation Average	2.32	3.28	3.66		
Maximum Deviation	8.7	7.9	11.9		
Statistical Significance	99.7%	99.1%	86.1%		

Evaluation The quality of the distances estimated by the application were evaluated in an experiment [AE08b]. The experiment took place in an environment without obstacles and the application was calibrated at distances of 40 meters and 100 meters. The experiment was conducted in three consecutive runs with the same set-up and the hardware used is listed within table 7.2. Table 7.3 and figure 7.8 shows the results of this experiment.

The highest accuracy occurs at points used for calibration. The lowest accuracy occurs at short distances below 30 meters (as expected). The average deviation is for all three runs below three meters and the maximum deviation was reached in the third experiment for the lowest possible distance (20m). Besides that the deviations were all below 10 meters. The measured values show a statistical significance between 86 and 99 percent according to the chi-square test in respect to the real node distances.

This shows, despite the limitations described above, that radio signal strength measurements can be used for distance estimation under specific conditions.

7.2.2. Topography based Radio Propagation Model

So far all experiments analyzed radio signal propagation on an open field in undeveloped areas without any obstacles influencing the signal propagation. These ideal environmental conditions are not very common in most real-life scenarios. Therefore we developed a topography based radio propagation model which also incorporates reflection and diffraction about obstacles (as described in section 4.3.2 on page 135).

In the following we present an evaluation of the topography based radio propagation model based on simulations as well as empirical data on the performance of the model. The theoretical evaluation of reflection and diffraction factors was performed in the Diplom thesis supervised by me of Steffen Reidt [Rei06] and the results were further elaborated and published in a joint paper [REKW11]. The validation by actual field measurements was performed within the Bachelor thesis supervised by me of Stefan Appel [App06] and the results were further elaborated and published in a joint paper [AE08a].

To evaluate the ray-optical model and to empirically assess the computational cost of different parameterizations, the model was implemented as an additional propagation model in the network simulator ns-2 [FV11]. Some additional data structures were required to store the topography model and the results of the ray casting algorithm.

7.2.2.1. Extensions of iNSpect

The ns-2 visualization tool iNSpect is extended in order to visualize the topography based radio propagation model. Figure 7.9 [REKW11] illustrates a new validation tab in iNSpect and the calculation results for an example scenario. The rays for direct line of sight, reflection and diffraction are shown in the second window while the configuration settings and resulting power values appear in the first window. Evaluation of computation results and determination of calculation periods has also been realized with the aid of this tool.

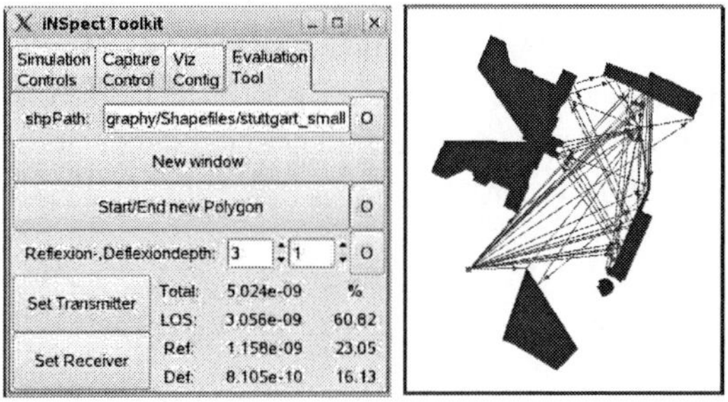

Figure 7.9.: iNSpect Validation Tab

The left part of figure 7.9 shows the a new tab "Evaluation Tool" which was added to the iNSpect user interface. It allows to specify a previously created shapefile that contains a topography model. New polygons can be interactively added to the simulation are by clicking on the window that represents the simulation field.

When a transmitter and a receiver are placed on the simulation field, the calculated radio propagation paths from transmitter to receiver are shown in the window. Furthermore, the absolute (milliwatts) and relative (percentage) shares of the different parts, direct line of sight, reflection and diffraction, transferred, are shown in respect to the total receiving power. Mixed paths that are composed of reflections and diffraction parts are counted as diffraction paths. The desired depth for reflection and diffraction calculation can be specified.

7.2.2.2. Reflection and Diffraction

Simulations were performed in order to test the robustness of the implementation of the radio propagation model and to estimate a reasonable calculation depth for reflection and diffraction. Evaluation of the results was performed numerically against existing models as well as experimentally. In addition visualization tools enabled qualitative validation as well as an assessment of relations between the maximum number of path interactions and calculation accuracy.

Reflection and Diffraction Depth In agreement with [Mau05] it was observed that consideration of paths with more than three reflections or one diffraction does not have an appreciable influence on power calculations. therefore the parameter settings shown in table 7.4 [Rei06] are recommended for the application of the topography based radio propagation model with respect to expected computation time in comparison to the simple free space propagation model.

Table 7.4.: Recommended Parameter Settings for Reflection and Diffraction Depth

Radio Propagation Model	Reflection Depth	Diffraction Depth	Time Factor
Free Space Propagation	-	-	1
Raycast	0	0	1,1 - 2,4
Raycast	2	0	12 - 40
Raycast	3	1	12 - 250

Reflection and diffraction depth specify the maximum number of reflections and diffraction that are considered for radio propagation paths. For very short calculation times reflection and diffraction depths of 0 can be used. In this case the connection between two nodes breaks down as soon as the line of sight is intersected by an obstacle. A compromise between calculation accuracy and calculation time, provides a reflection depth of 2 and diffraction depth of 0. The values were determined using various simulation scripts where the direct line of sight between nodes was interrupted for about 20% of the simulation time.

Given that calculation periods are highly dependent on the number of interactions that a path may contain, the maximum number of reflections and diffractions is configurable. Calculation periods for several simulation scenarios and configurations are discussed in the following.

Reflection and Diffraction Factors The receiving power is calculated as described in Section 4.3.2 by the length of the pathways and the reflection and diffraction factors with the help of the Two-Ray Ground model. We numerically validated the correctness of the results for reflection and diffraction factors, which are at the core of power calculations.

For these purposes we calculated the resulting *reflection factors* for horizontal and for vertical polarisation in dependency of the angle of incidence and *diffraction factors* for

- horizontal polarization and angle of incidence $\Phi = 45°$,
- vertical polarization and angle of incidence $\Phi = 45°$,
- horizontal polarization on a cone with interior angle of 90°, and
- vertical polarization on a cone of 90°.

The results for all evaluation scenarios are qualitatively consistent with data found in the literature for more elaborate models in [Mau05].

7.2.2.3. Computation Periods

All computations were performed on a Pentium Centrino 1.7 GHz processor with 1 GB of main memory. It should be noted however that the resources required for our model including shape file handling do not exceed 5 to 10 MB depending on the complexity and size of the terrain model. The host simulation environment ns-2 however requires considerable resources and therefore constrains the size and duration of the modeled period.

Figure 7.10 shows the calculation results for an example scenario that was evaluated using our iNSpect extension. This scenario shows a 600 m × 600 m square of the middle of London and contains 180 faces and 25 nodes. Such calculations are to be

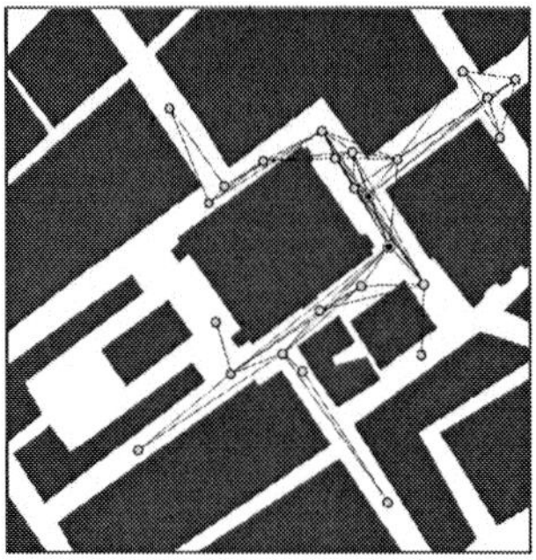

Figure 7.10.: Connectivity Between Nodes for Example Scenario.

performed on PDAs or other mobile resource-constrained devices to improve routing strategies.

Table 7.5.: Computation Times

Description	Time [s]
Direct connection of two single nodes	0.010
Multihop route with 3 hops	0.025
Multihop route with 5 hops	0.038
Multihop route with 7 hops	0.050
Connections between all nodes in Figure 7.10	0.60

Table 7.5 lists computation periods based on the scenario described above. Current handheld devices perform roughly at the same performance level as our test system. Improvements in equipment and ongoing optimization of our algorithms and implementation will further reduce the computation times exhibited by our proof of concept model. Results show that our proposed model is suitable for real-time use on resource-constrained (mobile) devices and can already be used effectively for cross data analysis to improve situation awareness.

7.2.2.4. Experiments with Obstacle

In the following the results are validated against actual field measurements based on two experiments we performed [AE08a]. In the following we compare the results of

Table 7.6.: Experimental Setups with Obstacle

Experiment 4	
Description:	Scenario with obstacle (reflection)
Environment:	Developed area
Line of sight:	√
Distance Measurement:	Measuring tape
Ground:	Stubble field
Experiment 5	
Description:	Scenario with obstacle (diffraction)
Environment:	Developed area
Line of sight:	–
Distance Measurement:	Measuring tape
Ground:	Stubble field

our model to measurements from two scenarios showing the most relevant effects (see table 7.6).

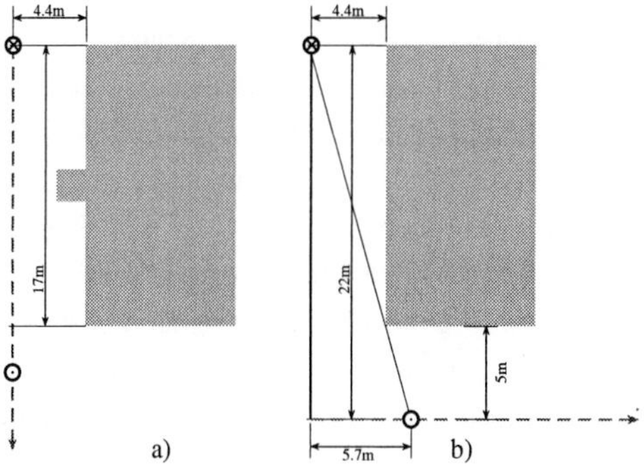

Figure 7.11.: Evaluation Setups with Obstacle: a) Experiment 4 (Reflection) and b) Experiment 5 (Diffraction).

Experiment 4: Reflection Scenario *Description* The first scenario in Figure 7.11 a) shows a building and two nodes, representing the sender (dotted circle) and the receiver (crossed circle). Both sender and transmitter are portable computers equipped with standard 802.11b/g network interfaces. While the sender has a fixed position, the receiver moves away following the dashed line in this scenario. It can be inferred from this figure that a direct line of sight existed during all measurements. Distance measurements were done using measuring tape.

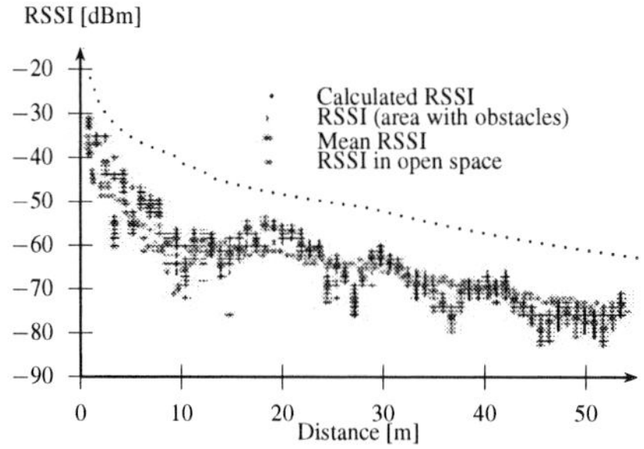

Figure 7.12.: Experiment 4: Scenario with Obstacle (Reflection)

Evaluation Figure 7.12 illustrates the corresponding power values of measurements and theoretical calculations for the first scenario. Black crosses represent measurements at intervals of 1 m and each square (illustrated in red) represents the mean power value for a specific distance. Stars (illustrated in blue) show results of measurements completed under identical conditions without any obstacles. Calculations, which were completed using the ray-optical propagation model, are given by a dashed line. Apart from the gap of 10 dBm described above, the curve of calculated values provides a good fit for measured values. Owing to simplifications in our model it is not possible to take interference effects into account, thus the curve of the calculated values shows a very smooth behavior whereas measurements show interference patterns most prominently caused by ground reflection (see also [Bal02]).

Power values provided by a network interface controller (NIC) are represented as RSSI (Received Signal Strength Indicator) [Bar02] therefore all power values had to first be measured in RSSI and converted to dBm. Unfortunately there is no standard for transforming RSSI into dBm or mW, so each card manufacturer defines the relation between RSSI and dBm as conforming to their own specification. This situation could explain an almost constant difference of 10 dBm between measured and calculated power values in Figure 7.12.

Another explanation for this measured difference could be the ground types in each simulation scenario. While the Two-Ray Ground model assumes level ground, the ground surface in the experiments was somewhat uneven and covered with vegetation. There is a gap of almost 10 dBm between measurements on concrete surfaces as opposed to grass due to the different permittivity of concrete and grass. While grass absorbs a high amount of transmitted power, concrete and similar surfaces are primarily reflecting transmitted power. As power reflected on ground surfaces yield interferences with power transmitted by line of sight, a poorly reflecting ground surface such as grass results in a higher overall receiving power.

Moreover transmitting power from the sender with a maximum transmitting power of 100 mW was not explicitly defined. Calculations were therefore based on a transmitting power of 100 mW which resulted in a 10 dBm gap; this gap does however highlight the desirability of choosing basic propagation parameters carefully and may indicate a need for incorporating ground permittivity in our constrained model.

Experiment 5: Diffraction Scenario *Description* Radio propagation in the first scenario was dominated by direct line of sight, building and ground reflection, the second scenario illustrates diffraction on a house wall (see figure 7.11 b)). While the sender has a fixed position the receiver is moving following the dashed line as before. This is disturbing the direct line of sight when the mobile notebook moves behind the building. Once again distance measurements were done using measuring tape.

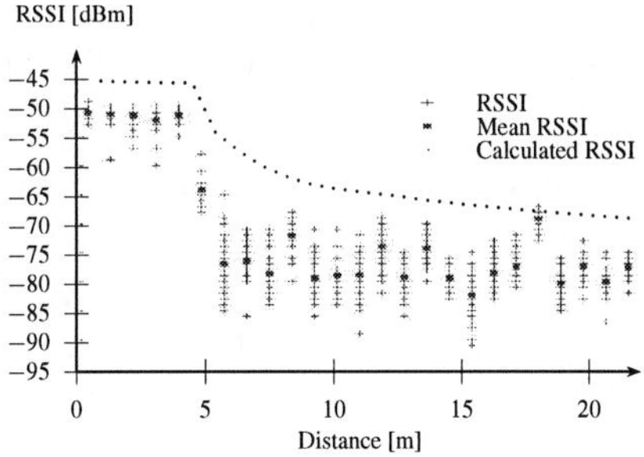

Figure 7.13.: Experiment 4: Scenario with Obstacle (Diffraction)

Evaluation As already shown in the first scenario the curve of calculated values shows a very smooth behavior and calculated values of our model show an very good general fit to measured values (see figure 7.13). After 5.7m the receiver loses its line of sight connection to the sender, resulting in a significant decrease of receiving power.

A different experiment examined the effect of a disturbed line of sight. The line of sight was interrupted by a wall at a distance of approximately 5 meters. This caused a dramatic decrease in signal strength, but as the values shown below, a connection between the two notebooks was still possible. Again the conversion of RSSI into dBm may explain an almost constant difference of 10 dBm between measured and calculated power values.

The results of these two experiments show that the surrounding environment has a high influence on signal propagation. Therefore signal propagation models must also consider obstacles around target nodes when estimating signal strengths for various distances.

Summary Some conclusions can be drawn from these experiments. In general, we observe a consistent behavior of real world experiment and the theoretical model. The propagation models presented in section 3.2.4.2 can be calibrated in accordance with experiments conducted in an environment without obstacles. Calibration is required for different hardware, orientation of the antennas, or the type of the surface used.

Overall results show conclusively that the approximation provided by our model is sufficient and suitable to reflect most significant physical effects such as reflection and diffraction about an obstacle. Evidence from investigated scenarios indicates a high degree of fidelity achieved by our constrained model compared to field measurements. We observed that parameters of the underlying free space and Two-Ray Ground models need to be chosen carefully and are investigating the incorporation of an additional permittivity factor for ground surfaces. Results also show that our proposed model is suitable for real-time use on resource-constrained mobile devices.

7.2.3. Location based Consistency Checks and Analysis

In this section we describe the implementation and evaluation of the attack detection proposed in section 4.3.3 on page 141. They provide consistency checks for location based data based on distance verification and connectivity examination [ES08]. The concepts were evaluated in various simulations using the JiST/MobNet [KBHS07] simulation environment which is based on the work of Rimon Barr [Bar04a]. Different set-ups were chosen to allow for a comprehensive analysis of the effectiveness of the proposed mechanisms [Som07].

MANETs of different sizes were simulated (size referring to both the simulation area, the number of nodes, and associated node density) and used for the evaluation (cf. [Som07]). The impact of node mobility and the presence of several attackers in the network were also investigated (cf. [EP09, PE09]).

Table 7.7 shows the parameters and the values that have been used in each simulation. The test series relate to the field size, node density, node mobility and the number of attackers.

Table 7.7.: Simulation Parameters for Different Evaluation Setups

Parameter	**Description**	**Values** *(* indicates default value)*
Field size	Size of the simulation area	300m×300m, 600m×600m, *900m×900m**, 1200m×1200m, 1500m×1500m
Number of nodes	Number of benign nodes in the simulation	4, 16, *36**, 64, 100
Mobility	Velocity of the nodes	no node movement, 0 to 1 m/s ($\approx 1.8\ \frac{km}{h}$), *1 to 2 m/s* ($\approx 5.4\ \frac{km}{h}$)*, 3 to 4 m/s ($\approx 12.6\ \frac{km}{h}$), 9 bis 10 m/s ($\approx 34.2\ \frac{km}{h}$)

In the evaluation cases related to the parameter field size, the field size as well as the number of nodes was changed in such a way that the node density is kept on the same level for all scenarios. An average number of 7 to 8 neighbors was chosen for each set-up to obtain a network with little fragmentation where multi-hop routing is possible and necessary. The calculations are based on the default radio range of 250 meters of the JiST/MobNet simulator.

For the evaluation of the influence of the node density on the performance, the parameter size remained unchanged at the standard value of 900m × 900m and only the number of nodes was changed.

7.2.3.1. Simulation Setups

The proposed attack detection methods were evaluated in various simulations using the JiST/MobNet [KBHS07] and the underlying routing protocol was Ad hoc On-demand Distance Vector Routing (AODV) [PBRD03].

For the evaluation of the attack detection system the most common attack scenario for MANETs, the black hole attack, was chosen. In this scenario the attacking node pretends to know short routes to other network nodes and announces them to its neighbor nodes, but it does not forge any positioning data or radio characteristics. Only two of the proposed analysis methods could therefore be evaluated since the black hole attack does not affect radio signal strength. In future work additional attack types should be modeled and implemented, e. g., wormhole attack, to allow a thorough analysis of all proposed mechanisms.

Table 7.7 shows the general simulation parameters and the values that have been used in each simulation. A total of four test series was performed for each analysis method, where one parameter was changed while other parameters were set to constant default values. The impact of the presence of several attackers in the network were also investigated (cf. table 7.7).

Each test series was performed 20 times to reduce statistical variance introduced by random generators, e. g., used for node placement, node movement, and data communication. The overall result derived is the average of individual results. Each simulation had a duration of 15 minutes (simulation time).

Table 7.8.: Number of Attackers for Different Simulation Setups

Parameter	**Description**	**Values** *(* indicates default value)*
Number of attackers	Number of black hole attackers	*1**, 2, 3, 5, 10

All test series were carried out with and without attacker to investigate the frequency of wrongly reported suspicions (false-positives) in different setups.

7.2.3.2. Evaluation Results

Each analysis method creates a *warning* when consistency checks indicate an anomaly, i. e. when one of the attack detection methods defined in section 4.3.4 on page 144 detects a conflict. A warning represents a weak suspicion that there might be an attack going on as apparent conflicts may arise also in normal network operation due to other reasons. For each analysis method the number of warnings during normal network operation and alternatively during an attack is compared to verify that the method can be used as a significant indicator for an attack.

All analysis methods wrongly report warnings even during normal network operation due to node mobility and the related dynamic changes of node positions and network topology. However, in case of an active attack the level of warnings should significantly rise to allow effective intrusion detection. The gap between normal network operation and attack is the indicator which can be used for attack detection.

Distance Verification The results of the distance verification are shown in figure 7.14 have two major characteristics.

For all test series the number of warnings in a scenario without an attacker is on average lower than in the same scenario with an attacker. This shows the influence of attacks on routing caused by incorrect routing information in relation to the actual distances between nodes. In figure 7.14(b) one can observe that with increasing node density the number of warnings *with an attacker* approaches the number of warnings *without attacker*. Therefore in very dense networks distance verification no longer supplies good evidence for an attack.

For the same number of nodes on a larger area the difference in the number of warnings increases significantly (see figure 7.14(a)). This reflects the fact that if the same number of nodes is located closer together within a relatively small area then an increased number of communication partners have a direct connection to each other. In this case routes between nodes are shorter and therefore a black hole attack can be more effective with far-distant routes. In contrast distance verification is very useful within

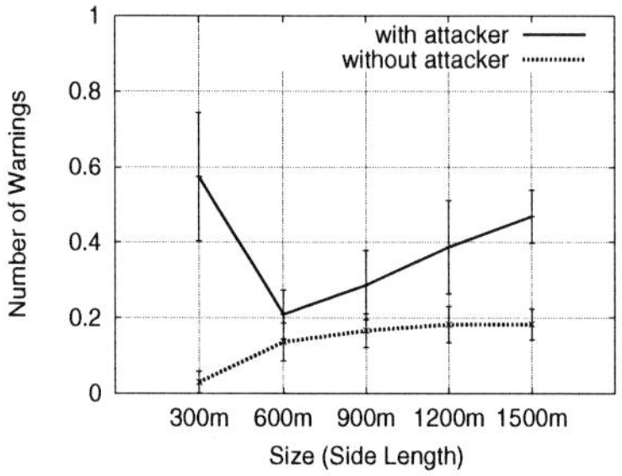

(a) Test series: size

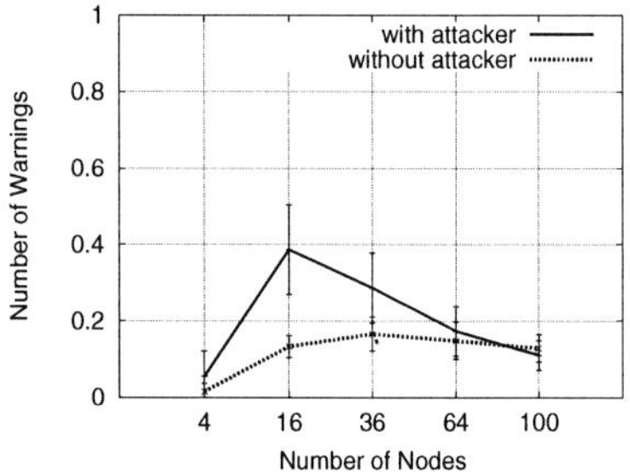

(b) Test series: node density

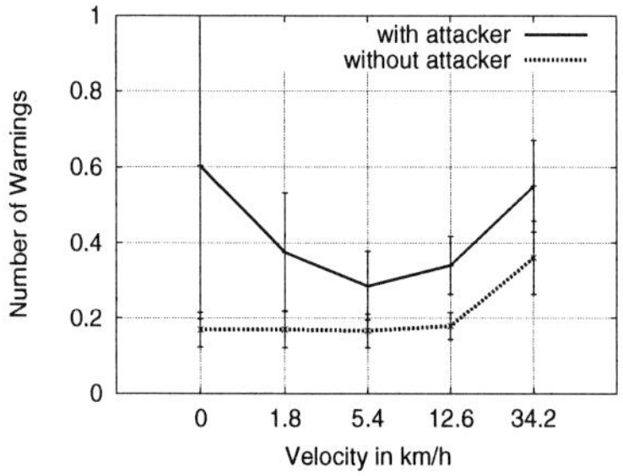

(c) Test series: mobility

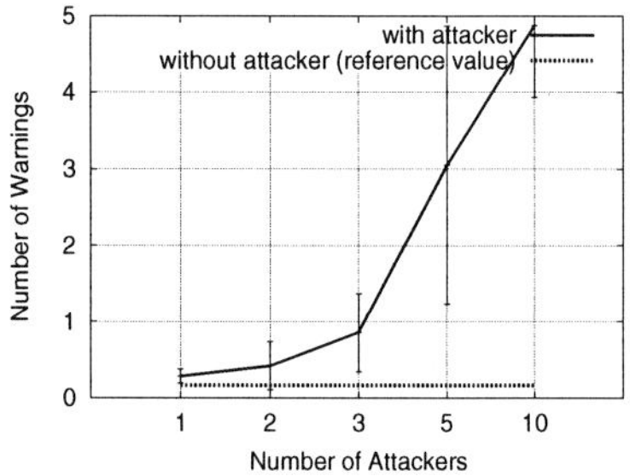

(d) Test series: number of attackers

Figure 7.14.: Relative Frequency of Warnings per Node and Analysis Process of the *Distance Verification* in Different Scenarios

networks with a small number of nodes, in which almost every route or route request is significantly influenced by the attacker (see figure 7.14(a)).

The second conclusion derived from the test results is the large variance of simulation results, as indicated in all charts by the vertical bars. Comparing the fluctuations with the results of other analytical methods, the differences become obvious. This behavior depends on the dynamic, but inaccurate determination of the maximum radio range. The location and connectivity of node relationship is less dependent on the scenario but rather on the position and the mobility patterns of nodes. Therefore, the fluctuations in the number of warnings that was encountered in all test series is particularly evident in the static network scenario without any node movements (see figure 7.14(c)).

The scale of the y axis in all figures of the test series with multiple attackers (see figure 7.14(d)) was adapted due to the large number of warnings. The number of warnings in a network without attacker is always shown as reference value for comparison. It can be clearly seen that an increasing number of attackers causes a significantly higher number of warnings.

Connectivity examination The connectivity examination shows similar behavior to the distance examination within small networks: the difference between the scenarios with and without attackers is significant (see figure 7.15(a)), since each node is directly affected by an attack. In the other scenarios the difference is smaller, but still significant. Similarly to the distance verification test, this analysis method shows a strong increase in the number of warnings for attackers with multiple attackers (see figure 7.15(d)).

The results for the test series regarding node density and mobility is similar to the distance verification. While with increasing mobility the number of warnings is increasing, the number of warnings per node decreases with increasing node density (see figure 7.15(b) and 7.15(c)). The difference between scenarios with and without an attacker is consistent and comparable in both test series. The increase in warnings with increasing node velocity originates from the continuous changes in network topology as previously known routes break down and new connections between nodes are established.

The decline of the number of warnings for denser networks is due to the increasing ratio of the total number of network nodes compared to the number of actually communicating nodes. In simulation setups with many nodes, the ratio of nodes that are involved in data communication is on average lower (presuming the same overall traffic load) and therefore the number of known routes between nodes decreases. Similarly less routing information is available for analysis and the relative frequency of warnings per node therefore decreases. This aspect also needs to be considered in the evaluation of the test series regarding field size, as with increasing field size the number of nodes in the network also increases (see table 7.7).

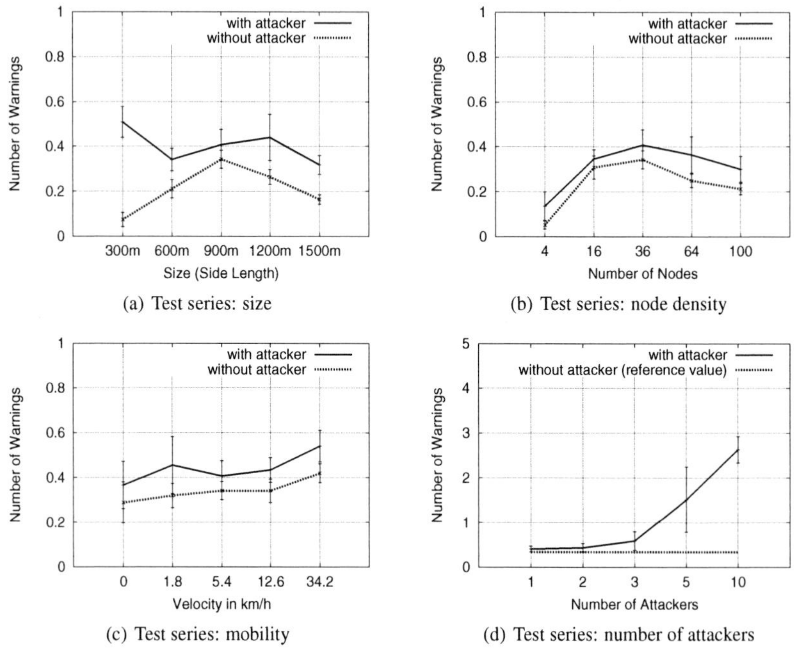

(a) Test series: size

(b) Test series: node density

(c) Test series: mobility

(d) Test series: number of attackers

Figure 7.15.: Relative Frequency of Warnings per Node and Analysis Process of the *Connectivity Examination* in Different Scenarios

7.3. Cooperative Trust Assessment: TEREC

In this section we evaluate the proposed cooperative trust assessment model TEREC presented in section 5.4 on page 171. The goal is the detection of malicious, non-cooperative nodes in order to improve the robustness of situation awareness. TEREC is evaluated in a MANET simulation environment as well as grid based mesh network. The evaluation was performed within the Master thesis supervised by me of Norbert Bißmeyer [Biß08] and the results were further elaborated and published in a joint paper [EB09].

In the first stage of the analysis several simulations were performed to optimize values of the weights. Resulting weights for calculating trust values are $w_{tc} = 0.8$ and $w_{tt} = 0.9$ as confidence values should have greater influence than trust values. Confidence values are calculated using the following weights: $w_{cc} = 1.2$ and $w_{ct} = 1.1$ which result in a higher impact of confidence values on calculation results. For the calculation of the trustworthiness the values $x = \sqrt{2}$ and $y = \sqrt{9}$ were chosen as proposed in [ZMHT05].

The *attacker model* for the evaluation allows Byzantine behavior of attacking nodes, i.e. they may deviate from normal behavior in an arbitrary way. In particular, instead of not contributing to the reputation system these nodes try to maliciously disturb accurate assessment and attempt to alter the system to an inconsistent state by inserting false values. Since we want to evaluate the worst case scenario for the cooperation mechanisms the attackers do not confine themselves to an inconspicuous behavior avoiding detection. Instead they try to attack and disrupt the system in the most effective way. Therefore malicious nodes continuously assign high trust and confidence values ($t = 1, c = 1$) to other malicious nodes and low trust and high confidence values ($t = 0, c = 1$) to benign neighbor nodes. This example behavior is chosen to demonstrate a worst case scenario where several attackers cooperate to attack the system. Benign nodes deliver high trust and confidence values for other benign nodes and low trust and high confidence values for malicious nodes.

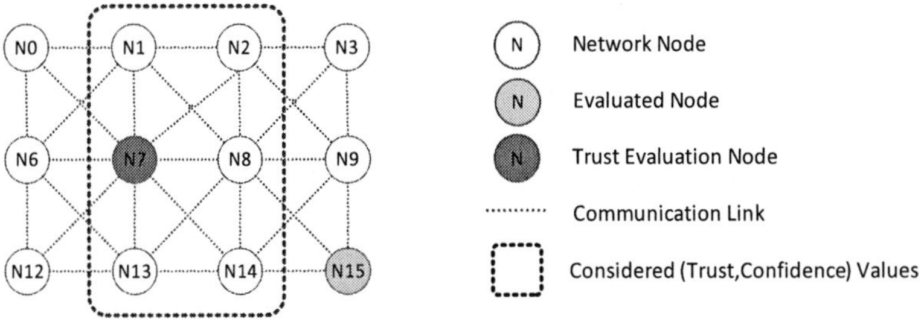

Figure 7.16.: Trust Values provided by Neighbours that are Considered on Updates

Trust and confidence values provided by neighbors with overly high hop counts above a certain threshold Δ_{hop_count} to the destination node are not incorporated. In figure 7.16 an example for an update progress with $\Delta_{hop\_count} = 1$ is shown. Node N6 evaluates the (trust, confidence) values for N12 and considers only (trust, confidence) values delivered from N2, N3, N7, N10 and N11. In this example the shortest route has a hop count of two hops, i. e., only nodes are considered where maximum route is one hop longer than the minimum (minimum hop count $+ \Delta_{hop\_count} = 2 + 1 = 3$). In every update the calculation of trust values for distant nodes (using equations 5.5 on page 177 and 5.6 on page 179) based on (trust, confidence) values provided by neighbor nodes is calculated and subsequently these (trust, confidence) values are combined as multiple assessments (using equations 5.7 on page 179 and 5.8 on page 181) to one pair of trust and confidence for each network node. If the node that is running the update has no active connection to the destination N12 a calculation of the shortest route is not possible and therefore the (trust, confidence) values from all neighbor nodes are considered.

7.3.1. Mobile Ad hoc Environment

For evaluation in a mobile environment the network simulator JiST/MobNet [KBHS07] was extended with the TEREC mechanisms. Several simulations are performed in a MANET scenario where the AODV [PBRD03] routing protocol is used to distribute reputation information piggybacking on normal routing packets.

20 nodes are placed on a simulation field of 1000m times 1000m. The radio range is set to 250m. As mobility model the random way point model is used with no pause time and the speed in between 1 and 2 m/s. One of the nodes is an attacker which can be detected by direct neighbor nodes.

Figure 7.17 shows detection of an attacker that starts malicious behavior after 100 rounds. As shown detection rate increases significantly following the first rounds of malicious behavior and almost all network participants (in the average 18 out 19) have detected the malicious node at the conclusion of the simulation.

Figure 7.18 shows the progress of trustworthiness of benign nodes N2, N8, N16 and the simulated malicious node N20 from the view of node N5. Because N5 is not a direct neighbor of node N20 in this case the detection of the bad behavior is delayed a few rounds, however it is important to note that the trustworthiness of the attacking node is constantly below the trusted value threshold as soon as node N20 commences malicious behavior. As a result an effective dissemination of information regarding

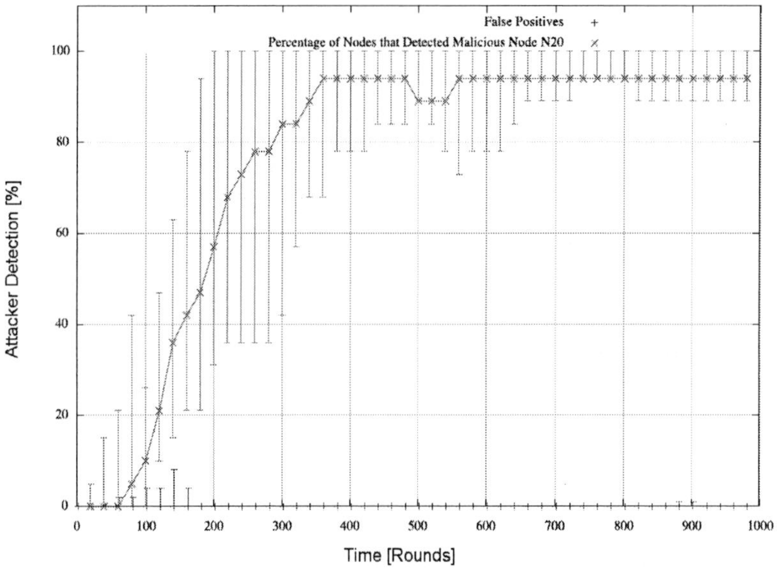

Figure 7.17.: Percentage of Detected a Simulated Attacker with AODV

malicious nodes is achieved; attacking nodes can be identified with 100% success rate with few false positives.

The following table shows a quantitative evaluation of the implementation in a random AODV network with a black hole attacker. The simulation was run 10 times in order to generate a mean value for various setups. In order to have a positive and stable environment at the start of the simulation the attacker only starts its malicious behavior after 100 rounds. Results presented in table 7.9 are therefore presented excluding the first 100 rounds of total simulation time.

7.3.2. Mesh Network

In order to evaluate the proposed concept in more detail under controlled conditions, a network grid without random mobility of nodes is used in the following. In two

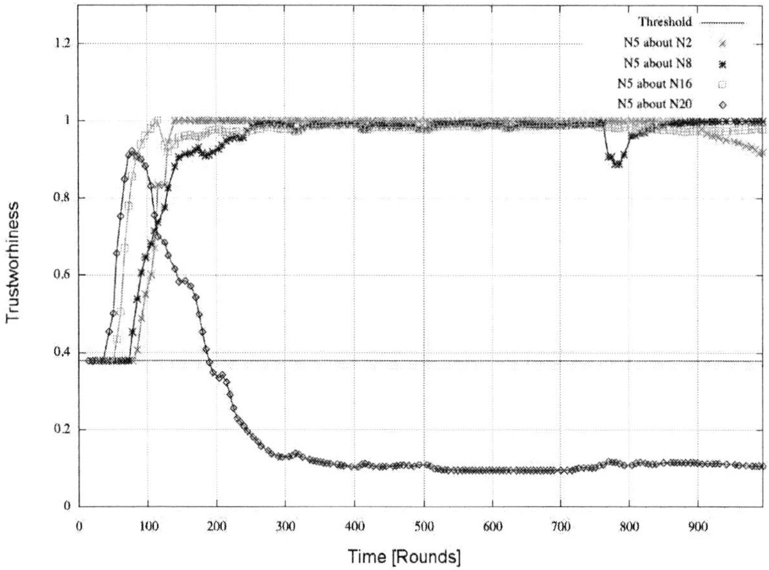

Figure 7.18.: Progress of Detecting a Simulated Attacker in an AODV based MANET

Table 7.9.: Time Until Attacker is Detected by a Certain Percentage of Nodes

Percentage	**Detection Time [Number of Rounds]**		
	Minimum	Mean	Maximum
50%	60	92	220
75%	120	142	240
100%	140	264	560

example configurations the trust system is evaluated and subsequently the distribution speed of reputation values is measured.

A network is created as a grid of 36 nodes. Every node non-border has a direct connection to eight neighbors. The system is round based meaning every cycle reputation values are disseminated and subsequently processed on every node.

Example with Two Malicious Nodes Figure 7.19 shows the setup of this example. There are two malicious nodes N21 and N28 that are acting benign in the first 25 rounds. Following the 25th round these nodes begin to behave maliciously.

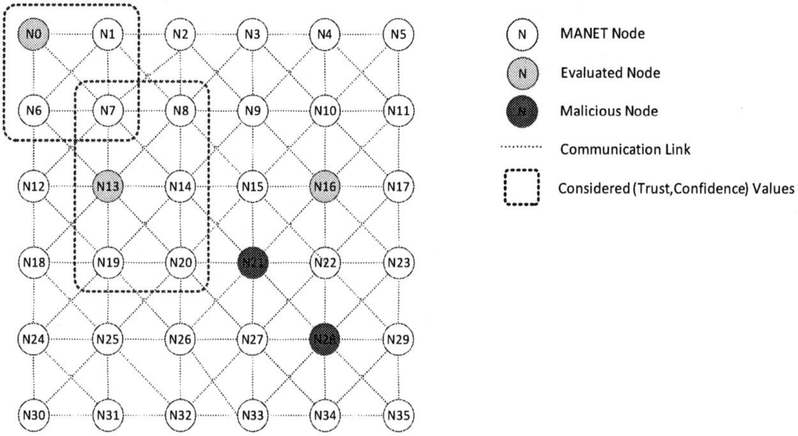

Figure 7.19.: Environment with Two Malicious Nodes

In this example reputation values from nodes with the smallest hop counts are accepted as well as values from neighbor nodes with one additional hop. This is critical for the demonstrated simulation as node N0 has only one route with minimum hop count to N28 passing through malicious node N21. If only this route would be used N0 would have a positive default reputation from N28 and not a negative value. As a solution to this problem the radius of consideration for neighbor nodes is expanded to the minimal hop count plus one (as shown in figure 7.19 – the grey area behind node N0 and N13). As N22 is a direct neighbor of node N28 it does not consider any additional information about this adjacent node from other neighbor nodes.

Figure 7.20 shows the progression of trustworthiness value estimations of compromised node N28 within the network from the viewpoint of three example nodes N0, N13 and N22. The diagram shows trustworthiness values from 100 rounds in the example network. In the first 25 rounds all nodes increase their trustworthiness regarding node N28 and following round 25 the direct neighbors of node N28 rapidly decrease their trustworthiness. Nodes with a hop count greater than two need a longer period of

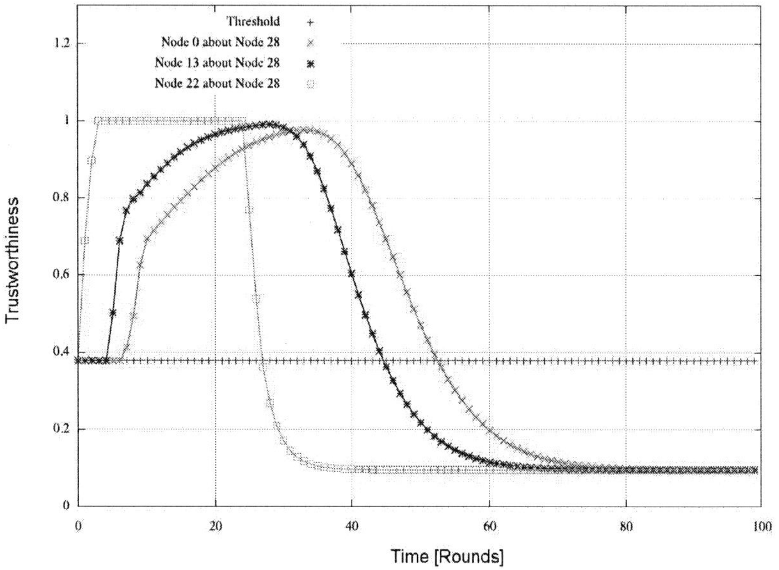

Figure 7.20.: Trust Progress of Node N0, N13, N22 about Node N28 with Two Malicious Nodes

time until they receive an unfavourable trustworthiness value for the malicious nodes. Although most routes from node N0 and N13 go through the second malicious node N21, accurate behavior values are still transmitted over the network.

Example with Eleven Malicious Nodes In the second example eleven malicious nodes are added to the grid network. As shown in figure 7.21 malicious nodes are distributed over the entire grid with a majority in the proximity of node N13.

All nodes are evaluating node 28 which has decreasing trustworthiness values after round 25 similar to the first example. Figure 7.22 shows the progress of trustworthiness values for node 28 from the viewpoint of node N0, N13 and N22. In contrast to the previous example node N13 has five malicious nodes in its direct neighborhood. In this scenario node N13 detects all its direct malicious neighbors N8, N14, N19 and N20.

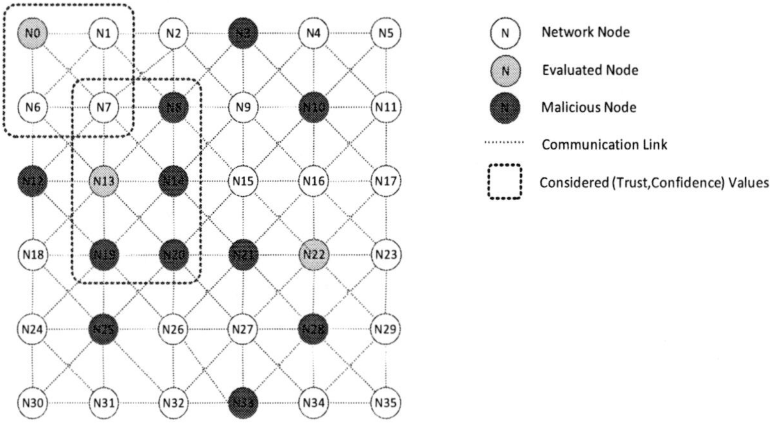

Figure 7.21.: Environment with Eleven Malicious Nodes

Subsequent calculations do not consider the reputation values of these nodes. The only trustworthy node for N13 that is relevant with respect to node N28 is N7, however the reputations of N7 are manipulated by malicious neighbor nodes in every round. As a result the malicious behavior of node N28 is not detected by N13.

Node N0 has in this example enough trustworthy neighbor nodes to accurately evaluate the trustworthiness of node N28. As shown in figure 7.22 the number of rounds until N28 has reached the threshold has increased in comparison to the first example in section 7.3.2. Due to the high number of malicious nodes in the direct neighborhood the malicious behavior of node N28 is propagated slowly.

Example with High Number of Malicious Nodes In the third example there is a high number of malicious nodes present in the neighborhood of node N0 which preclude an accurate evaluation N28. For this test the same network is used as shown in figure 7.21 but with node N9 as a malicious node. The majority of nodes in relation to N28 is malicious and therefore provides positive values for node N28. Node N0 therefore gets incorrect information from trustworthy benign nodes in the one-hop neighborhood as all paths to node N28 are blocked by malicious nodes. As result N0

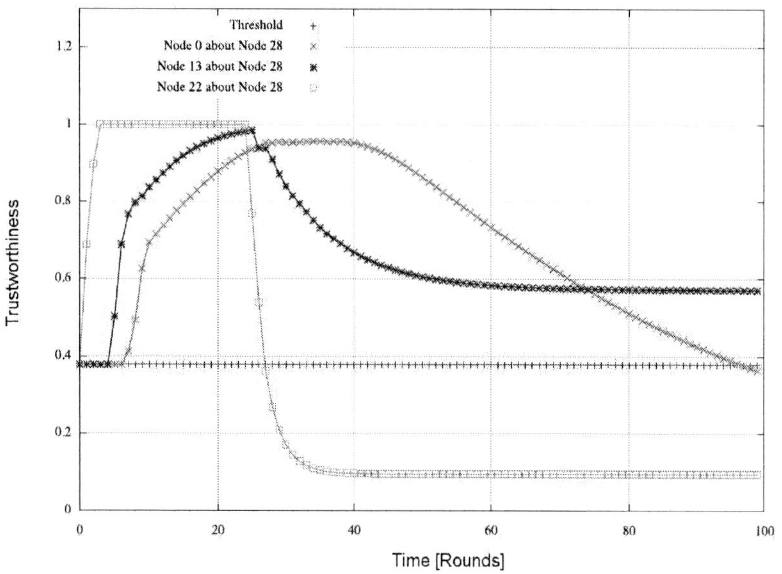

Figure 7.22.: Trust Progress of Node N0, N13, N22 about Node N28 with Eleven Malicious Nodes

and N13 have trustworthy values for the malicious node N28. Consequently for the proposed reputation system it is not only important how many nodes are malicious but also how malicious nodes are situated in the network.

7.3.3. Quantitative Evaluation of the Trust System

In contrast to the examples provided in the previous sections an overview of the network with *all possible* attacker node position is given in the following example. This section provides a quantitative evaluation of a network with 36 nodes where all nodes are organized into a grid. In general 36 different evaluations have to be performed with one simulated attacker in each round. The attacker is placed on every possible position of the network (N0, N1, N2 ... N35) and subsequently evaluated to determine how many iterations it takes for all other nodes to detect this malicious node. The behavior of the malicious node is identical to that described in the previous trust system evaluation 7.3.2.

Table 7.10.: Number of Rounds Until Attacker is Detected by a Certain Percentage of Nodes

Attacker Node ID	50%	75%	100%	Maximum Distance
14, 15, 20, 21	*13*	*18*	*21*	3
7, 10, 25, 28	15	21	28	4
8, 9, 13, 16, 19, 22, 26, 27	13	18	25	4
0, 5, 30, 35	22	28	35	5
1, 4, 6, 11, 24, 29, 31, 35	20	25	32	5
2, 3, 12, 17, 18, 23, 32, 33	18	23	29	5
Average	16.9	22.1	28.4	

In table 7.10 evaluation results with variable attackers are shown. The round is recorded when 50%, 75% and 100% of all nodes in the network have detected the attacker. In order to test a more realistic environment all nodes (including malicious nodes) start with maximum positive reputation base values prior to malicious nodes subsequently starting their negative behavior.

The first row in the table shows the shortest detection times. If the attacker is node N14, N15, N20 or N21 the maximum distance between the attacker and all other nodes in the network has a hop count of 3. In this case the detection time is 21 rounds and it raises approximately proportional to the maximum distance between the attacker and the other network nodes. For nodes with a maximum distance of 4 hops the detection time is between 25 and 28 rounds. If node N0, N5, N30 or N35 are acting as attackers the maximum distance between the nodes has a hop count of 5 and the detection time is 35 rounds.

7.4. Probabilistic State Modeling: Task Force Tracking

In this section we evaluate the PF based Task Force Tracking approach (SMC-TFT) as an example for probabilistic state modeling and estimation [EKW12]. The newly proposed method incorporating additional information sources (topographic data and mobility model) is compared with two standard TFT mechanisms. We first present the simulation setup and then discuss the evaluation results.

7.4.1. Simulation Setup

The basic setup is that each mobile node contains a computing device, has GPS sensor capability and topography model and mission information are available on each node. Nodes regularly exchange data (geo-location, velocity) to direct neighbors via multi-hop flooding with all other nodes (e. g., piggyback with routing information). Location measurement and message processing are performed in discrete time steps (round based simulation) where the duration of each step was set to 1 s.

In the following sections we describe the default parameter set used for our evaluation. It is explicitly mentioned in the description whenever default values are not used.

7.4.1.1. Basic Parameters

The default setup contains 20 nodes that move within a simulation area of 1000 m x 1000 m based on one of the mobility models described in the following section. Nodes locally determine their location and velocity using the GPS sensor. The GPS measurement standard deviation for location is $\sigma_{GPS}^{loc} = 3.0$ m and for velocity $\sigma_{GPS}^{vel} = 1.0$ m/s. Every 10 s each node generates a location message and forwards it to all its neighbor nodes.

Each simulation is carried out for 250 s (i. e., 250 rounds) of which the first 50 s are left out as initialization and stabilization period and therefore are not considered for the evaluation. Each simulation setup is run in 50 iterations to determine stable average values. The number of particles for the SMC-TFT approach is 50.

7.4.1.2. Mobility Model

For the evaluation we implemented a mobility model reflecting the topography of the environment and a group based model reflecting tactical formations (cf. section 3.2.4.1 on page 64).

Manhattan Grid Model The *Manhattan Grid Model [Spe98]* is based on road topology. Roads are located in a grid structure and nodes move in horizontal or vertical directions on these roads. The model follows a probabilistic approach: each node probabilistically chooses at each intersection to keep moving in the same direction or to turn left or right.

For the evaluation we implemented the Manhattan Grid Model based on the BonnMotion [AEGPS10] algorithm. This model is used as default mobility model for the evaluation. There are 10 blocks in each direction (i. e., each block is 100 m x 100 m) and the probability for mobile nodes to turn at a crossing is 0.5. The probability for a node to change its speed (in a simulation round) is 0.2. The minimum speed is 0.5 m/s, the mean 2.0 m/s and the standard deviation of a normally distributed random speed 2.0 m/s. The pause probability is zero.

Reference Point Group Mobility (RPGM) Model The *Reference Point Group Mobility (RPGM)* Model [HGPC99, CBD02] can represent tactical relationships among a group of mobile nodes. For the evaluation the following default parameter set is used. There is a group of 20 nodes that move with a common average speed. The minimum group speed is 4 m/s and the maximum 10 m/s. The overall group movement follows a random waypoint mobility model. Each node chooses a destination randomly distributed around a group destination in a distance of up to 25 m from this specified location. In each step a node may also randomly move up to 30 percent of this maximum distance.

7.4.1.3. Radio Propagation Model

During evaluation a deterministic and a probabilistic radio propagation model are used (cf. section 3.2.4.2 on page 68).

Free Space Model The Free Space Model is a basic deterministic model where only line-of-sight radio transmissions through free space (without any obstacles, reflection or diffraction disturbances) are taken into account. For the evaluation we use the available simulator implementation [HE04].

Shadowing Model In reality radio propagation is a probabilistic process due to multipath propagation effects. The shadowing model consists of two distinct parts: a path loss model (which predicts the mean received power) and a second that reflects the variation of the received power at a certain distance.

For the evaluation an implementation based on ns-2 [FV11] is used. This model is used as default radio propagation model with a radio transmission range of 250 m. The pathloss exponent β is set to 4.0 (recommendation for outdoor environment/shadowed urban area: 2.7 to 5 [FV11]) and a shadowing deviation $\sigma_{dB} = 8.0$ is used (recommendation for outdoor environments: 4 to 12 [FV11]).

7.4.2. Evaluation Results

We compare our newly proposed SMC-TFT method with two standard TFT mechanisms: A simple straight forward TFT mechanism, which simply takes a measurement "as is" and a more elaborate mechanism based on dead reckoning (cf. [AD03]) considering the progress in time between last measurements and current time as well as node velocity.

Standard TFT (STD-TFT) The most recently received measurement $\mathbf{z}_l$ (GPS coordinates and velocity) is taken as an estimate of current location and velocity of a node:

$$\mathbf{x}_k = \mathbf{z}_l$$

Velocity Based TFT (VEL-TFT) The location accuracy is improved by predicting the current location based on available measurement data (cf. [AB09]). Previous node location, time difference $\Delta t = t_k - t_l$ between measurement time t_l and current time t_k as well as node velocity are considered:

$$\mathbf{x}_k^{pos} = \mathbf{z}_l^{pos} + \Delta t \cdot \mathbf{z}_l^{vel}$$
$$\mathbf{x}_k^{vel} = \mathbf{z}_l^{vel}$$

7.4.2.1. Accuracy of Location Estimation

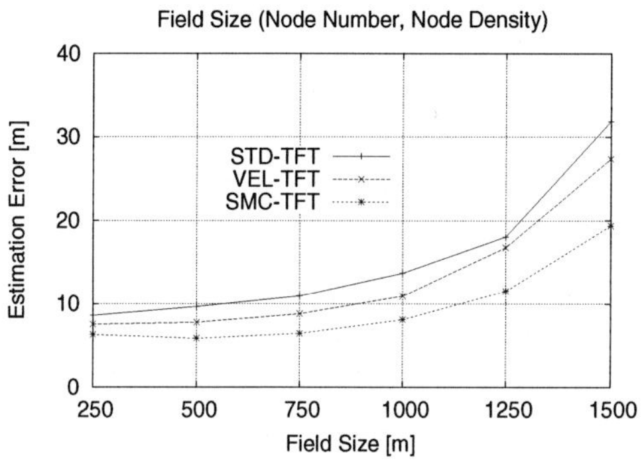

Figure 7.23.: Average Estimation Error vs. Node Density for Manhattan Grid Model

Field Size, Node Number and Node Density In this evaluation setup we increase the field size (incrementally from 250 m x 250 m to 1500 m x 1500 m) proportional to the number of nodes (5 to 30). Therefore node density decreases with increasing field size and node numbers. Figure 7.23 shows that the estimation error stays on a similar level as long as node density provides some basic connectivity of all nodes, but decreases when network connectivity is disturbed. The SMC-TFT method outperforms other methods in all setups by 16 to 41 percent.

Mobility Model Parameters In figure 7.24 and figure 7.25 simulation results are shown for different speed settings for Manhattan Grid and RPGM models respectively. The results for both mobility models show that STD-TFT works well for scenarios with low mobility.

In figure 7.24 the minimum speed for the Manhattan Grid Model is set to 0.5 m/s for speeds below 3 m/s and to 1 m/s otherwise. Mean speed as well as the standard deviation are increased from 1 m/s to 4 m/s. Estimation errors grow relatively proportional to speed. Average estimation errors of SMC-TFT are 23 to 48 percent lower than for both standard TFT methods.

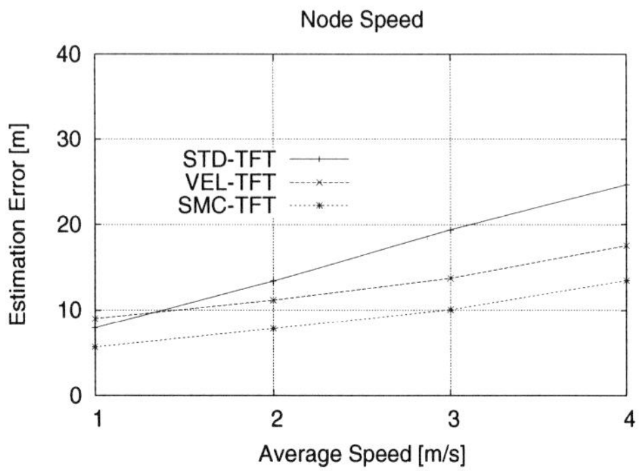

Figure 7.24.: Average Estimation Error vs. Node Speed for Manhattan Grid Model

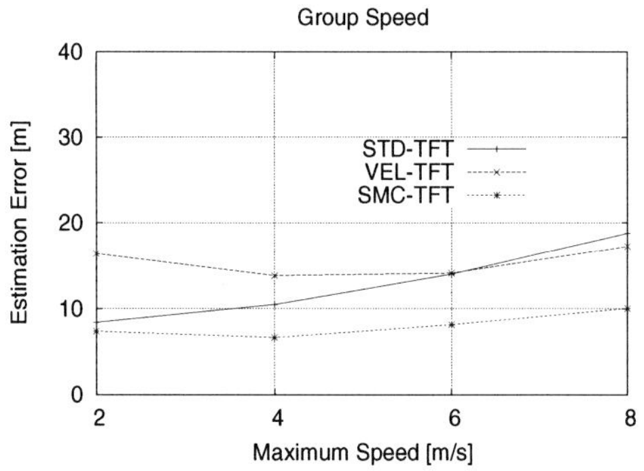

Figure 7.25.: Average Estimation Error vs. Group Speed for RPGM Model

In figure 7.25 maximum group speed of RPGM is increased from 2 m/s to 8 m/s, minimum group speed is increased proportionally from 0.5 m/s to 2 m/s. Average estimation errors of SMC-TFT are 12 to 55 percent lower than for the other methods.

Radio Propagation Model Parameters The following setup evaluates the radio propagation models influence on TFT performance. For these purposes two radio propa-

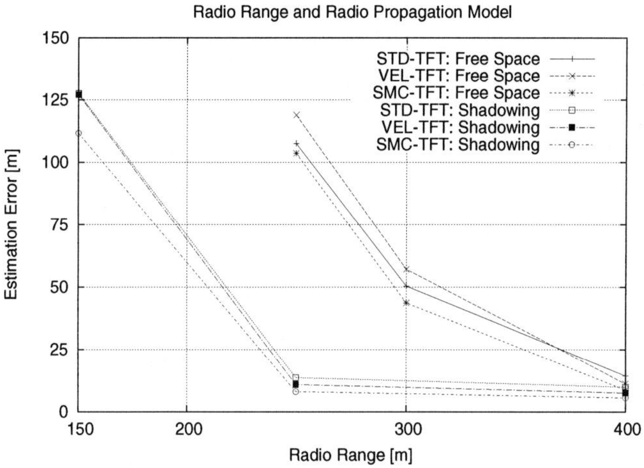

Figure 7.26.: Radio Propagation Model and Radio Range

gation models (Free Space and Shadowing) are used with varying radio ranges from 150 m to 400 m where radio range is the maximum transmission range for the Free Space model and the average transmission range for the Shadowing model. Figure 7.26 shows that error estimation is significantly lower for the Shadowing model. When even a few messages are transmitted beyond "normal" (Free Space) radio range it significantly improves location estimation due to an increase in valuable information that would otherwise not be available to that part of the network. The results show that the average estimation error is lower for SMC-TFT for all evaluated setups than for both standard TFT methods.

GPS Measurement Accuracy Accuracy of sensor measurements has a direct impact on TFT performance. Figure 7.27 shows results for varying GPS precisions (GPS standard deviation 2 m to 5 m). Estimation error increases are relatively proportional to GPS accuracy where the estimation error of the proposed SMC-TFT is 25 to 38 percent than for the other methods.

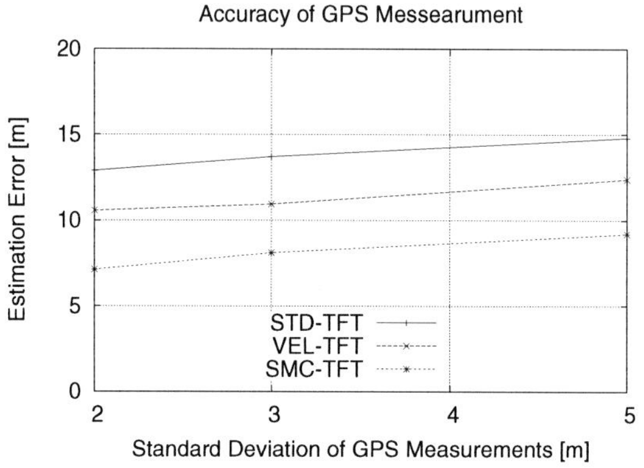

Figure 7.27.: GPS Mesurement Accuracy

7.4.2.2. Communication Overhead

Table 7.11.: Number of Messages and Average Estimation Error vs. Message Generation Interval

	Message Generation Interval		
	5 s	**10 s**	**20 s**
Average Number of Generated Messages	481	240	120
Average Number of Forwarded Messages	33512	16576	8488
Average Estimation Error			
STD-TFT	8.4 m	13.7 m	24.5 m
VEL-TFT	6.7 m	10.9 m	20.9 m
SMC-TFT	4.6 m	8.1 m	15.0 m
Reduction (by SMC-TFT)	31.3 %	25.7 %	28.2 %

In this section we evaluate TFT communication overhead and the resulting location estimation errors in dependency of message generation intervals. As message generation, forwarding and processing do not depend on the data fusion algorithm, the number of messages that are generated and exchanged are equivalent for all three methods.

Table 7.11 shows that the quantity of messages significantly decreases when location message generation time intervals are increased. Similarly location estimation error increases proportionally to a decrease in message generation intervals where SMC-TFT outperforms the other approaches in all scenarios by more than 25 percent.

7.5. Summary

In this chapter we described the implementation and evaluation of example applications for all three main contribution areas: cross data analysis, cooperative trust assessment and probabilistic state modeling. We specified for each of them various experimental setups covering a large range of application scenarios and evaluated the proposed solutions for these setups.

Cross Data Analysis: Location Related Information We evaluated the application of the cross data analysis approach for location related information. For this purpose we analyzed the interrelationships between node distances and radio characteristics based on radio signal strength measurements in various environmental setups [REKW11,AE08a]. This was the basis for the evaluation of the consistency checks for location based attack detection [ES08].

A series of experiments was conducted to examine the suitability of signal strength for distance estimation. Measurement analysis showed that radio signal strength can in principle be used for the estimation of node distances. The applicability depends very much on the surrounding environment. In settings with obstacles, particular in urban environments, a model of the environment would have to be incorporated with a more complex radio propagation model. In order not to be limited to certain WLAN cards or surrounding conditions the application should be calibrated for a specific set-up [AE08a].

We developed a topography-based radio propagation model which incorporates reflection and diffraction about obstacles [REKW11]. We evaluated the reflection and diffraction factors of the ray-optical model. The theoretical simulations were validated experimentally to conclusively show that the approximation provided by our model is sufficient and suitable to reflect the most important physical effects such as reflection and diffraction about an obstacle. Simulation results show that our proposed model is suitable for real-time use on resource-constrained mobile devices.

We evaluated the attack detection approach based on location related information which exploits interdependencies of positioning data, radio signal characteristics and topology information [ES08]. The consistency check mechanisms were evaluated with respect to their accuracy and applicability for attack detection using the simulation environment JiST/MobNet for various MANET scenarios. The evaluation results always shows a clear difference for MANETs during normal operation compared to during an attack. This proves that the evaluation of data that is directly or indirectly related to node localization and underlying network topology is a powerful approach for the detection of active attacks which is an important component for situation awareness in tactical MANETs.

Cooperative Trust Assessment: TEREC The concept of using (trust, confidence) values in order to evaluate node behavior was implemented in the TEREC system and evaluated for situation awareness in tactical MANETs in order to identify malicious, non-cooperative nodes [EB09]. TEREC was evaluated in a MANET simulation environment as well as grid based mesh networks. The results show that the proposed method is also robust in simulation environments with variable number of attackers. A quantitative evaluation of reputation value dissemination speed was presented.

Probabilistic State Modeling: Task Force Tracking We evaluated the Task Force Tracking example application [EKW12] for the probabilistic state modeling and estimation concept [EW09]. The results show that Task Force Tracking can be significantly enhanced by using networking capabilities of tactical MANETs and state estimation based on particle filters incorporating external information such as topographic data or mission information (e. g., mobility models). As shown in the simulation environment using various mobility and radio propagation characteristics (cf. Figure 7.23 to 7.26), our proposed SMC-TFT model enhanced both accuracy and robustness as compared to existing models.

The analysis presented in this chapter exemplifies how the proposed concepts can be applied. It is presented how the cross data analysis of location related information can be exploited in order to detect active attacks. TEREC provides robust mechanisms for cooperative trust assessment in order to identify malicious nodes. Furthermore it is shown that probabilistic state modeling is a good basis for the incorporation of additional information sources in order to increase the performance of Task Force Tracking in tactical MANETs. Overall the evaluation results show how the proposed concepts can increase robustness of situation awareness in tactical MANETs.

8. Summary and Outlook

The objective of our research presented in this dissertation is to improve situation awareness in tactical mobile ad hoc networks (MANETs). Tactical teams are supported to successfully accomplish their missions. As described in chapter 1 on page 3 (*Introduction*) this could be a team of first responders after a natural catastrophe or a terrorist attack. For example, following the Haiti earthquake in 2010 rescue forces provided technical and humanitarian aid for the affected population. Small teams tried to find victims and provide food and water supplies in difficult to access areas on the island.

In this kind of scenarios the situation is dynamically evolving and robust and effective mechanisms are required to enable task forces to quickly respond to recent developments. It is crucial that they are aware of the current situation, e. g., location and status of task forces and victims, environmental situation, other resources and entities such as harmful substances. This information is crucial for mission success as it enables informed decisions and can save lives. The existing infrastructure may be destroyed (or should not be used for other reasons) and a backup communication infrastructure needs to be set up. Mobile devices connected by a MANET are a quick, flexible and efficient way to provide such an infrastructure.

Existing situation awareness solutions for tactical MANETs are not sufficiently protected against faults and security threats (as discussed in section 1.1.3 on page 7). The proposed approaches for data processing and situation assessment are limited in dealing with wrong (malicious or faulty) information. They process different kinds of data separately and do not collectively analyze and exploit correlated or redundant data sources. Potential conflicts within data sets due to missing or faulty information may be resolved but other available data sources are not considered in order to detect malicious behavior. They only consider cross-layer information of the communication system and do not incorporated additional information sources [KLTM10, KLAML11].

In distributed environments as tactical MANETs collaborative aspects are important for the cooperative information assessment and trustworthiness of remote information sources. In particular, characteristics of wireless communication should be exploited, for example, trust information could be distributed to all direct neighbor nodes at the same time [TB04, BPPR09]. The temporal and causal relationships and related problems in distributed state estimation are not sufficiently considered and resolved [OUR*05, OBDWU08].

In chapter 2 on page 17 (*Problem Definition*) we discussed the problem addressed within this dissertation on a technical level. We presented typical application areas for tactical MANETs in emergency response and disaster recovery and characteristics of situation awareness in this context. Then we described the main objectives and research challenges. As node mobility and related aspects pose particular challenges in MANET environments, location-related aspects are a main focus within this dissertation. The overall objective is increasing robustness in state estimation for situation awareness in tactical MANETs. This objective was broken down into three research challenges represented by the following three questions:

1. How can location-related *data sources* be utilized in a *comprehensive and effective way* to detect malicious behavior?
2. How can the *trustworthiness* of other nodes be *efficiently and cooperatively* assessed?
3. How can *system states* be *modeled in a generally applicable way* and *comprehensively assessed* to counter inaccurate and error-prone data?

In chapter 3 (*Related Work*) we analyzed the state of the art related to the three research challenges addressed within this dissertation. We presented a review of related work, relevant concepts and definitions proposed in the literature.

In the following section we summarize our main contributions in respect to the three identified research challenges described in part II: chapter 4 (*Cross Data Analysis*), chapter 5 (*Cooperative Trust Assessment*) and chapter 6 (*Probabilistic State Modeling*).

Then we conclude this dissertation by giving an outlook on research topics that could be further examined and elaborated in future work.

8.1. Summary of Contributions

The main focus of this work is on improving state estimation in tactical MANETs. This is the basis for increasing robustness of situation awareness, e. g., in respect to error-prone and/or malicious data. In the following we present the proposed concepts addressing the three research challenges in more details.

(1) How can location-related *data sources* be utilized in a *comprehensive and effective way* to detect malicious behavior? We developed a general concept for *cross data analysis* in order to increase robustness in situation awareness in tactical MANETs (see chapter 4 on page 115). The presented model provides a concept for exploiting all available data sources in order to detect malicious behavior. The proposed concept incorporates *direct and indirect sensor data and additional knowledge sources* (such as mission-specific knowledge and general information sources) in order to detect inconsistencies [EKW12].

For this purpose we analyzed and specified overlaps and cross-relationships between data sources. For detecting overlapping data sets the temporal and spatial correlation of specific data elements as well as the subject that they refer to need to be considered. *Multiple data sources are analyzed collectively* and cross-relationships are exploited to detect malicious data.

We investigated what kind of cross-relationships may be exploited for cross data analysis and categorized them based on their nature. We derived definitions of consistency, inconsistency and contradictions within this context. The *explicit modeling of cross-relationships* within our approach enables the derivation of consistency checks. Consistency checks are utilized to discover contradictions across multiple data sets in order to detect false or malicious data [ES08]. The developed concept increases robustness in situation awareness in tactical MANETs as inconsistencies are not only detected within data provided by one sources but also across multiple data sources.

The developed concept for *cross data analysis was applied for location related data sources*. Therefore the available data sources related to node location available in tactical MANETs and their cross-relationships were analyzed and modeled.

An efficient topography based radio propagation model was developed and implemented in order to *exploit a topography model of the environment as an additional knowledge source*. This way *radio signal strength measurements were incorporated as indirect sensor data* for distance estimation [AE08a]. The topography based radio propagation model allows to model and analyze cross-relationships between network topology and node positions [REKW11].

Node locations, routing information, radio signal characteristics and a topography model of the environment are *collectively modeled and analyzed based on the cross data analysis concept* to detect malicious behavior. Based on this combination of multiple location related data sources algorithms for the detection of active attacks in tactical MANETs were developed and implemented as a proof of concept [ES08].

The cross data analysis concepts were *evaluated based on field trials and simulations* (see section 7.2 on page 222). Field trials in various environmental setups showed that radio signal strength can in principle be used for the estimation of node distances and that the applicability strongly depends on the application scenario. The topography based radio propagation model was implemented as an extension of the network simulator ns-2 and the visualization tool iNSpect. The results of the simulative evaluation of the topography based radio propagation model are for all evaluation scenarios are qualitatively consistent with data found in the literature. The implemented model is suitable for real-time use on resource-constrained mobile devices.

The location based consistency checks and attack detection mechanisms were evaluated with respect to their accuracy and applicability. For this purpose simulations were carried out for various MANET scenarios using JiST/MobNet. The considered attacks focus on the communication infrastructure, attackers try to compromise the routing behavior in order to fake a wrong network topology. The results show that data that is directly or indirectly related to node localization and underlying network topology is a powerful approach for the detection of active attack. The attack detection mechanisms always show clear difference of MANETs during an attack in comparison to normal operation.

The successful validation of this proof of concept shows that cross data analysis can be utilized to increase robustness of situation awareness in tactical MANETs. Contradictions in data sets can be more effectively detected exploiting cross-relationships between data sources. This improves the detection capabilities and increases detection rates. In dynamic MANET environments this helps to cope with wrong data due to sensor faults as well as with malicious behavior of attacking nodes. Tactical teams get a clear and correct picture of the situation and are able to successfully complete their missions.

(2) How can the *trustworthiness* of other nodes be *efficiently and cooperatively* assessed? We developed efficient and robust mechanisms for cooperative trust assessment in tactical MANETs (see chapter 5 on page 151). The focus of the proposed concepts is on the assessment of the trustworthiness of other network nodes which provide information for situation awareness. For this purpose, we *combine efficient broadcast gossiping trust aggregation* and *trust information modeling as (trust, confidence) value pairs* [EB09]. The presented gossiping based trust assessment architecture exploits the wireless medium by distributing trust information to all direct neighbor nodes at the same time. An efficient and simple information local trust aggregation mechanism limits the amount of data that is flooded into the network. Therefore the communication and processing overhead is significantly reduced.

The effectiveness of the trust assessment mechanisms is further enhanced considering the trust level of the providing node when *incorporating second hand trust information.* For these purposes the trust value is complemented with a confidence value representing the certainty about a specific trust assessment for a specific network node. The trust modeling as (trust, confidence) value pairs enables the mitigation of spurious ratings by checking the congruence of views and lowering the confidence when incorporating contradiction assessments.

We developed and specified an *instantiation of the proposed concepts called TEREC* [EB09]. For this instantiation we described and specified the mechanisms and protocols in detail. This includes the mechanisms for local trust assessment and the gossiping based exchange with direct neighbor nodes. We specified the trust value calculation for indirect assessments as well as for the combination of multiple opinions and a trust assessment protocol for the overall process.

The proposed TEREC model was implemented and *evaluated based on simulations* in various scenarios (see section 7.3 on page 246). Simulations based on JiST/MobNet were used to evaluate the dynamics of the model over time in a mobile ad hoc environment. The performance of the system when an increasing number of nodes behaves malicious was analyzed based on a grid based mesh network setup. The evaluation results show that the system is very robust system and that trust assessments quickly converge to the correct value even in environments with several attackers.

The proposed trust assessment scheme allows a benign majority of nodes to prevail and accurately classify network nodes based on observations and trust estimations. The cooperative trust assessment mechanisms improve the resistance of the situation awareness to attackers. Network nodes responsible for wrong or malicious behavior can be identified and excluded from the situation awareness process. This can ensure the success of a tactical mission even in the case of attack.

(3) How can *system states* be *modeled in a generally applicable way* and *comprehensively assessed* to counter inaccurate and error-prone data? We developed a concept for *probabilistic state modeling and estimation* for situation awareness in tactical MANETs (see chapter 6 on page 185) to increase the overall robustness regarding inaccurate and error-prone input data [EW09]. The probabilistic modeling of system states provides increased flexibility and robustness as estimations are not restricted to a specific value but instead are represented by a "probability cloud".

The proposed *concept based on particle filters* is open and *applicable to a wide range of application scenarios* (including nonlinear state propagation as well as Non-Gaussian process noise). The particle filter based state estimation process allows *incorporating of additional information at multiple stages.* Additionally, the *observation history is incorporated* estimating state probabilities over entire paths and do not only considering the current state.

Within the proposed concept we *incorporate multiple observations and additional information sources* [EKW12], e. g., mission information, domain-specific background knowledge and general information sources. These information sources may pose some constraints on the state estimation process and are exploited to adjust the likelihood of specific system states or exclude impossible states. The proposed concept allows the efficient modeling and incorporation of constraints and additional information sources in order to increase the robustness of the state estimation process.

We presented a *distributed state estimation architecture* [EW09] which addresses the challenges resulting from the distributed nature of tactical MANETs. Suitable observations and state estimates are selected and exchanged within tactical MANETs for improved cooperative state estimation. We also considered and resolved temporal and causal relationships of distributed particle filters as information received from other network nodes may refer to a moment in time in the past. The re-simulation mechanism resolves these correlation of observations by erasing part of a particle filter's history and running it forward again to the current time incorporating all collected observations at appropriate intervals.

The concepts were applied to a specific application scenario: *Task Force Tracking (TFT)* [EKW12]. This is a typical *example application for tactical situation awareness.* The presented mechanism incorporates additional information such as mission information (e. g., mobility models) and topographic data. It is contains the major building block described above serves as a proof of concept for our probabilistic state modeling approach based on particle filters.

The TFT model based on particle filters and additional information was implemented and evaluated (see section 7.4 on page 255). A simulation environment was used to evaluate the model for various MANET scenarios using different mobility and radio propagation models. The results show that both accuracy and robustness are enhanced using the proposed model compared to standard TFT models.

Probabilistic state modeling increases the robustness of the state estimation process to measurement noise. State estimation based on particle filters can flexibly be applied for basically any kind of relevant situation information. Incorporating additional information sources improves the preciseness of state estimation based on error-prone data sources. State estimation results are more accurate and the situation awareness picture is more realistic. This way tactical task forces can be supported as good as possible even in case of attack in order to successfully complete their mission.

8.2. Future Research

In the following we present related and future research topics as well as potential extensions of the proposed concepts. Firstly, it could be further elaborated how the developed concepts can be *integrated within the whole situation awareness cycle* including situation projection (level 3), decision making and the implementation of actions. Some of the presented concepts may also be applied and extended for these components to further enhance the robustness of situation awareness in tactical MANETs. Secondly, the interaction and integration of all three major areas of contributions could be analyzed. Probabilistic state modeling was already combined with cross data analysis as mission information and additional knowledge sources were incorporated in the Task Force Tracking concept.

In some scenarios the ad hoc based communication infrastructure may also be *complemented with a fixed infrastructure or (intermittent) back-links* to a back-end infrastructure or *unmanned aerial vehicles (UAVs)* may be used to increase the availability or the bandwidth of the local MANET. Therefore it could be investigated how the developed concepts can be applied and adjusted for hybrid scenarios. Additionally, *coalition environments* are quite common in tactical MANETs. Multiple groups and organizations, such as THW and Red Cross, may join their resources to communicate and collaborate for a specific task or mission. The developed concepts could be analyzed in respect to the specific requirements and challenges that may arise in coalition environments.

The *cross data analysis concept* could be *applied to other application types* and respective situation awareness information in tactical MANETs. This may show the benefits of the proposed concepts and further increase the robustness of situation awareness. The application of the cross data analysis in other application areas could also provide further insights and may be exploited to further elaborate the cross data analysis concept. It would be interesting to *conduct further field trials using the topography based radio propagation model*. This way it may be determined how well and reliable it performs under various conditions and applications scenarios in order to know the applicability and limitations for situation assessment.

The *theoretical analysis of the developed cooperative trust assessment model* may be further elaborated. It could be even investigated if mathematical models could be developed that analytically describe the trust assessment process in comparison to other models (at least on some abstraction level) or provide some general constraints on the information diffusion speed.

The *strategies for selection and exchange of information within the distributed particle filter architecture* could be elaborated for additional application scenarios. The developed concepts could exploit the expected adversary models or the availability of additional information sources, e.ġ. mission procedures and objectives. The *efficiency of the presented re-simulation process* may be significantly improved. An analysis of the benefits of re-simulation depending on the surrounding conditions may help to reduce the number of re-simulation steps while keeping the expected benefit on a high level.

The proposed probabilistic state modeling concept could be further elaborated for *hybrid state estimation*. The individual steps of the particle filter based state estimation process could be analyzed for further extension or optimization when discrete states are included.

Finally, it could be further analyzed and elaborated how existing or new concepts (e. g., in multi-object tracking) can be applied and extended to *follow multiple hypothesis in parallel* during situation assessment. A single or a group of faulty sensors may lead to a consistent track of observations (and respective observed states) that do not fit to the general set of observations. An attacker may try to lead the situation awareness system on a wrong track. In both cases following multiple tracks in parallel allows to gather additional information and to take an informed decision.

Appendices

A. Publications and Talks

The dissertation is partially based on the following publications and talks:

A.1. Publications

1. EBINGER, P., KUIJPER, A. AND WOLTHUSEN, S. D. Enhanced Location Tracking for Tactical MANETs based on Particle Filters and Additional Information Sources. In *Proceedings of Military Communications and Information Systems Conference. MCC 2012* (2012), IEEE.
2. REIDT S., EBINGER, P., KUIJPER, A. AND WOLTHUSEN, S. D. Resource-Constrained Signal Propagation Modeling for Tactical Mobile Ad Hoc Networks. In *Proceedings of the IEEE 1st International Worksthop on Network Science. NSW* (2011), IEEE.
3. EBINGER, P., AND WOLTHUSEN, S. D. Efficient State Estimation and Byzantine Behavior Identification in Tactical MANETs. In *Proceedings of the IEEE Military Communications Conference. MILCOM 2009* (2009), IEEE.
4. EBINGER, P., AND PARSONS, M. J. Measuring the Impact of Attacks on the Performance of Mobile Ad hoc Networks. In *Proceedings of the Sixth ACM International Symposium on Performance Evaluation of Wireless Ad-Hoc, Sensor, and Ubiquitous Networks* (2009), ACM.
5. PARSONS, M. J., AND EBINGER, P. Performance Evaluation of the Impact of Attacks on Mobile Ad hoc Networks. In *Proceedings of Workshops Dependable Network Computing and Mobile Systems* (2009).
6. EBINGER, P., AND BISSMEYER, N. TEREC: Trust Evaluation and Reputation Exchange for Cooperative Intrusion Detection in MANETs. In *IEEE Seventh*

Annual Conference on Communication Networks and Services Research (2009), IEEE.

7. JAHNKE, M., WENZEL, A., KLEIN, G., EBINGER, P., ASCHENBRUCK, N., GERHARDS-PADILLA, E., AND KARSCH, S. MITE – MANET Intrusion Detection for Tactical Environments. In *NATO Research & Technology Organisation (RTO) IST-22-PBM Symposium* (2008). Received Best Paper Award.
8. APPEL, S., AND EBINGER, P. Entfernungsschätzungen basierend auf Funksignalstärkemessungen für die Angriffserkennung in MANETs. In *D-A-CH Security 2008. Proceedings* (2008), syssec.
9. EBINGER, P., AND SOMMER, M. Using Localization Information for Attack Detection in Mobile Ad hoc Networks. In *Sicherheit 2008* (2008), GI LNCS.
10. EBINGER, P., AND BUCHER, T. Modelling and Analysis of Attacks on the MANET Routing in AODV. In *Proceedings of 5th International Conference on AD-HOC Networks & Wireless (AdHoc-NOW)* (Ottawa, Canada, August 2006), T. Kunz and S. S. Ravi, Eds., vol. 4104 of *Lecture Notes in Computer Science*, Springer.
11. BUCHER, T., EBINGER, P., TÖLLE, J., AND JAHNKE, M. Modellierung und Analyse von Angriffen auf das MANET-Routing in OLSR. In *D-A-CH Mobility 2006: Bestandsaufnahme, Konzepte, Anwendungen, Perspektiven* (Ottobrunn, Germany, October 2006), P. Horster, Ed., syssec, pp. 156–172.

A.2. Talks

1. *Enhanced Location Tracking for Tactical MANETs based on Particle Filters and Additional Information Sources*
   Military Communications and Information Systems Conference (MCC), Gdansk, Poland. 9.10.2012.
2. *Efficient State Estimation and Byzantine Behavior Identification in Tactical Mobile Ad hoc Networks (MANETs)*
   IEEE MILCOM 2009, Boston, USA. 19.10.2009.

3. *Energy-Efficient Key Distribution and Revocation in Tactical Networks with Asymmetric Links*
IEEE MILCOM 2009, Boston, USA. 19.10.2009.
4. *Performance Evaluation of the Impact of Attacks on Mobile Ad hoc Networks*
Workshops Dependable Network Computing and Mobile Systems, Niagara Falls, USA. 27.9.2009.
5. *TEREC: Trust Evaluation and Reputation Exchange for Cooperative Intrusion Detection in MANETs*
IEEE Seventh Annual Conference on Communication Networks and Services Research, Moncton, Canada. 13.05.2009.
6. *Entfernungsschätzungen basierend auf Funksignalstärkemessungen für die Angriffserkennung in MANETs*
D-A-CH Security 2008, Berlin, Germany. 25.6.2008.
7. *Using Localization Information for Attack Detection in Mobile Ad hoc Networks*
GI Sicherheit 2008, Saarbrücken, Germany. 4.4.2008.
8. *Modellierung und Analyse von Angriffen auf das MANET-Routing in OLSR*
D-A-CH Mobility 2006, Ottobrunn, Germany. 18.10.2006.
9. *Modelling and Analysis of Attacks on the MANET Routing in AODV*
5th International Conference on AD-HOC Networks & Wireless (AdHoc-NOW), Ottawa, Canada. 18.08.2006.
10. *Sicherheit in mobilen Ad-hoc-Netzen*
Informatik-Kolloquium (Computer Science Colloquium), University of Jena, Germany. 26.06.2006.

A.3. Other Publications

1. VASILOMANOLAKIS, E., FISCHER, M., MÜHLHÄUSER, M., EBINGER, P., KIKIRAS, P., SCHMERL, S. Collaborative Intrusion Detection in Smart Energy Grids. *1st International Symposium for ICS & SCADA Cyber Security (ICS-CSR 2013)* (2013).
2. EBINGER, P., HERNÁNDEZ RAMOS J.L., KIKIRAS, P., LISCHKA, M., WIESMAIER, A. Privacy in Smart Metering Ecosystems. *1st EIT ICL Labs Workshop on Smart Grid Security (SmartGridSec 2012)* (2012).
3. SCHINZEL, S., SCHMUCKER, M., EBINGER, P. Security Mechanisms of a Legal Peer-To-Peer File Sharing System. *IADIS International Journal on Computer Science and Information Systems 4* (2009), pp. 59-72.
4. EBINGER P., CASTRO NEVES M., SALAMON R. AND BAUSINGER, O. Challenges for the Implementation and Revision of International Biometric Standards Demonstrated by the Example of Face Image Data. In *Proceedings of the Special Interest Group on Biometrics and Electronic Signatures conference. BIOSIG 2009.* (2009), A. Brömme, Ed., Gesellschaft für Informatik (GI).
5. LAUTENSCHLÄGER, F., AND EBINGER, P. Using XSL Transformations for Java Code Generation of ASN.1 Data Structures. In *Proceedings of the IADIS International Conference Applied Computing 2008* (Algarve, Portugal, April 2008), pp. 19–26.
6. EBINGER, P., CASTRO NEVES, M., SALAMON, R., AND SEIBERT, H. International Database of Facial Images for Performance and ISO/IEC 19794-5 Conformance Tests. In *BIOSIG 2008. Proceedings* (Darmstadt, Germany, September 2008), pp. 165–174.
7. APPEL, S., AND EBINGER, P. *Lokalisierung in mobilen Ad-hoc-Netzen ohne zusätzliche Infrastruktur: Entfernungsschätzungen und Topologiebestimmung.* VDM Verlag Dr. Müller, Saarbrücken, 2008.
8. EBINGER, P., SCHINZEL, S., AND SCHMUCKER, M. Security Mechanisms of a Legal Peer-To-Peer File Sharing System. In *Proceedings of the IADIS International Conference Applied Computing 2008* (Algarve, Portugal, April 2008), pp. 86–93.

9. NG, K. ; DANG, M. ; ONG, B. ; EBINGER, P. ; OPEL, A. ; CAMPO, R. AXMEDIS Programme and Publication Tools for Cross Media Content Multi-channel Distribution. In *Proceedings of the Fourth International Conference on Automated Production of Cross Media Content for Multi-Channel Distribution* (AXMEDIS 2008), 2008, pp. 261-265.

10. ULZHEIMER, J., SCHMUCKER, M., AND EBINGER, P. CONFUOCO on BitTorrent – Liable File Sharing with Swarming. In *Proceedings of the Second International Conference on Automated Production of Cross Media Content for Multi-Channel Distribution* (AXMEDIS 2006), K. Ng, Ed., vol. Volume for Workshops, Applications and Industrial, University of Florence, Firenze University Press, pp. 196–203.

11. EBINGER, P., AND PINSDORF, U. Internetdetektive auf Patrouille. *KEM 6* (June 2006), 60–61. S. 60-61, Konradin Verlag.

12. NG, K., ONG, B., NEAGLE, R., EBINGER, P., SCHMUCKER, M., BRUNO, I., AND NESI, P. AXMEDIS Framework for Programme and Publication and On-Demand Production. In *Proceedings of 1st International Conference on Automated Production of Cross Media Content for Multi-channel Distribution* (Florence, Italy, November 2005).

13. PINSDORF, U., AND EBINGER, P. Automated Discovery of Brand Piracy on the Internet. In *International Conference on Parallel and Distributed Systems Workshops. Proceedings Volume 2: ICPADS-2005 Workshops* (Fukuoka, Japan, July 2005), J. Ma and L. T. Yang, Eds., IEEE Computer Society, pp. 550–554.

14. SCHMUCKER, M., AND EBINGER, P. Alternative Distribution Models based on P2P. In *Proceedings of 3rd International Workshop for Technical, Economic and Legal Aspects of Business Models for Virtual Goods* (Florence, Italy, November 2005).

15. SCHMUCKER, M., AND EBINGER, P. Promotional and Commercial Content Distribution based on a Legal and Trusted P2P Framework. In *7th International IEEE Conference on E-Commerce Technology 2005* (2005), IEEE CEC 2005.

16. EBINGER, P., AND PINSDORF, U. Automatisierte Recherche von Markenpiraterie im Internet. In *D-A-CH Security 2005: Bestandsaufnahme, Konzepte, Anwendungen, Perspektiven. Gemeinsame Arbeitskonferenz GI · OCG · BITKOM ·*

SI · TeleTrusT (Darmstadt, Germany, März 2005), P. Horster, Ed., syssec, pp. 250–263.

17. EBINGER, P., AND JALALI-SOHI, M. PKI-Unterstützung für mobile Endgeräte durch Server-Delegierung. In *IT Sicherheit im verteilten Chaos, 8. Deutscher IT-Sicherheitskongress des BSI* (Bonn, Germany, May 2003), SecuMedia Verlag, Ingelheim, Germany, pp. 281–293.
18. JALALI-SOHI, M., AND EBINGER, P. Towards Efficient PKIs for Restricted Mobile Devices. In *Proc. IASTED Int. Conf. Communications and Computer Networks* (Calgary, Canada, November 2002), M. Hamza, Ed., Acta Press, pp. 42–47.
19. EBINGER, P., AND TESKE, E. Factoring of $N = pq^2$ with the Elliptic Curve Method. In *In Proceedings of ANTS V (Sydney, Australia, July 2002), LNCS 2369* (2002), Springer, pp. 475–490.
20. NOCHTA, Z., EBINGER, P., AND ABECK, S. PAMINA: A Certificate Based Privilege Management System. In *In proc. of Network and Distributed System Security Symposium (NDSS02)* (San Diego, USA, February 2002).

B. Supervising Activities

The following list summarizes the student bachelor, diploma and master theses supervised by the author. The results of these works were partially used as an input into this dissertation.

B.1. Diploma and Master Theses

1. *Angriffserkennung und -prävention in mobilen Ad-hoc-Netzen basierend auf einem transparenten Routing-Proxy*, Fatih Gey
   Diploma Thesis, Technische Universität Darmstadt, 1.5.2009–31.10.2009
2. *Entwicklung von Bewegungsmodellen für die realitätsnahe Simulation mobiler Ad-hoc-Netzwerke*, Philipp Bormuth
   Master's Thesis, Hochschule Furtwangen, 15.8.2008–15.2.2009
3. *Distributed Data Collection and Analysis for Attack Detection in Mobile Ad hoc Networks*, Norbert Bißmeyer
   Master's Thesis, FH Joanneum, Graz, Österreich, 1.4.2008–15.9.2008
   Awards: GRAWE High Potential Award 2008, CAST Award IT Security 2009
4. *Evaluation and Extension of a Detection System for Attacks on MANET Routing*, Javier Saez de Arregui
   Diploma Thesis, Mondragon Unibertsitatea, Spanien, 1.4.2008–15.9.2008
5. *Implementierung und Evaluierung des MANET-Routingprotokolls OLSRv2 auf Basis der Simulationsumgebung JiST / MobNet*, Carsten Eichelberger
   Diploma Thesis, University of Applied Sciences Gießen-Friedberg, 1.1.2008–30.6.2008

6. *Konzeption und Implementierung von Angreifermodellen und verbesserten Methoden für die Angriffserkennung in mobilen Ad-hoc-Netzen*, Stefan Endler
Master's Thesis, Technische Universität Darmstadt, 1.11.2007–30.4.2008
7. *Auswertung von Positionsdaten, Funkcharakteristika und des Verbindungsgraphen für die Angriffserkennung in mobilen Ad-hoc-Netzen*, Martin Sommer
Diploma Thesis, Technische Universität Darmstadt, 1.4–30.9.2007
8. *Signaturbasierte Erkennung von Angriffen in mobilen Ad-hoc-Netzen*, Tobias Combé
Diploma Thesis, Technische Universität Darmstadt, 1.10.2006–31.5.2007
9. *Security in Mobile Ad Hoc Networks*, Eneko Ibarzabal Atutxa Diploma Thesis, Mondragon Unibertsitatea, 1.11.2005–30.6.2006
10. *Topographisches Routing in mobilen Ad-hoc-Netzen*, Steffen Reidt
Diploma Thesis, Technische Universität Darmstadt, 1.1.2006–30.6.2006
11. *Modellierung und Analyse von Angriffen auf Routing-Verfahren in mobilen Ad-hoc-Netzen*, Tobias Bucher
Diploma Thesis, Technische Universität Darmstadt, 21.9.2005–21.12.2005

B.2. Bachelor Theses

1. *Evaluation der Auswirkungen verschiedener Angriffstypen in mobilen Ad-hoc-Netzen*, Malcolm Parsons
Bachelor's Thesis, Technische Universität Darmstadt, 15.8.2008–15.11.2008
Award: CAST Award IT Security 2009
2. *Lokalisierung von Knoten in mobilen Ad-hoc-Netzen ohne zusätzliche Infrastruktur*, Stefan Appel
Bachelor's Thesis, Technische Universität Darmstadt, 15.6.2006–15.9.2006
3. *Anpassung des Routingprotokolls Ad hoc On-Demand Distance Vector Routing für Composite Routing Metrics*, Wael Ghouweish
Bachelor's Thesis, Darmstadt University of Applied Sciences, 1.4.2006–31.10.2006

B.3. Other Supervised Theses

1. *Entwicklung eines legalen Peer-to-Peer Filesharingsystems auf der Basis des BitTorrent-Protokolls*, Jochen Ulzheimer
   Master's Thesis, Darmstadt University of Applied Sciences, 6.4.2006–6.10.2006
2. *Sicherheitsmechanismen eines legalen Filesharingsystems*, Sebastian Schinzel
   Bachelor's Thesis, Darmstadt University of Applied Sciences, 15.7.2005–15.10.2005
3. *Benutzerverwaltung und Client eines legales Filesharingsystems*, Johannes Stehlik
   Diploma Thesis, University of Applied Sciences Gießen-Friedberg, 1.4.2005–30.9.2005
4. *Peer-to-Peer-Kommunikation und Content-Management eines legalen Filesharingsystems*, Ferruh Zamangör
   Master's Thesis, Darmstadt University of Applied Sciences, 1.4.2005–30.9.2005
5. *Automatisierte Recherche von Markenpiraterie im Internet: Semantische Analyse und Bewertung*, Jens Vogel
   Diploma Thesis, University of Applied Sciences Zittau/Görlitz, 1.3.2005–31.8.2005
   Award: CAST Award IT Security 2005
6. *Automatisierte Recherche von Markenpiraterie im Internet: Informationsbeschaffung und -vorverarbeitung*, Simon Buß
   Diploma Thesis, University of Applied Sciences Osnabrück, 1.3.2005–31.8.2005
   Award: CAST Award IT Security 2005
7. *Konzeption und Implementierung eines Codegenerators für ASN.1-Datenstrukturen*, Frank Lautenschläger
   Studienarbeit, Technische Universität Darmstadt, 1.6.2004–30.11.2004

C. Curriculum Vitae

Personal Data

Name	Peter Ebinger
Birth date & place	July 17, 1973 in Backnang, Germany
Nationality	German

Education

Oct. 1996 – Nov. 2001	Diploma in Computer Science, University of Karlsruhe, Germany (grade: sehr gut/A) Thesis topic: "Design and Prototypical Implementation of an Infrastructure for Certificate based Privilege Management"
Sept. 1999 – Jun. 2000	Exchange student, University of Waterloo, Canada
Oct. 1994 – Sept. 1996	Vordiplom in Physics, University of Karlsruhe (grade: gut/B)

Work Experience

Oct. 2011 – present	Senior Researcher, Security Research Group, AGT Group (R&D) GmbH, Darmstadt, Germany
May 2008 – Sept. 2011	Technical Head, Evaluation Laboratory, Fraunhofer IGD, Darmstadt
Jan. 2008 – Sept. 2011	Deputy Head, Competence Center Identification and Biometrics, Fraunhofer IGD
May 2002 – Sept. 2011	Research Associate, Security Technology Department (subsequently Competence Center Identification and Biometrics), Fraunhofer IGD
Mar. – Apr. 2002	Visiting researcher, University of Waterloo
May 1997 – Jul. 1999	T.A. of Prof. P. Schmitt, University of Karlsruhe, Creation of algorithm animations for computer science lecture notes

Bibliography

[AB09] AMAR E., BOUMERDASSI S.: Enhancing Location Services with Prediction. In *Proceedings of the 2009 International Conference on Wireless Communications and Mobile Computing* (2009), ACM, pp. 1025–1029.

[AC02] ALBERS P., CAMP O.: Security in Ad Hoc Networks: A General Intrusion Detection Architecture Enhancing Trust Based Approaches. In *Proceedings of the First International Workshop on Wireless Information Systems, 4th International Conference on Enterprise Information Systems* (2002).

[AD03] AGARWAL A., DAS S. R.: Dead Reckoning in Mobile Ad Hoc Networks. In *Proceedings of the IEEE Wireless Communications and Networking Conference (WCNC)* (2003), vol. 3, pp. 1838–1843.

[AE08a] APPEL S., EBINGER P.: Entfernungsschätzungen basierend auf Funksignalstärkemessungen für die Angriffserkennung in MANETs. In *Proceedings of the D-A-CH Security 2008* (2008), syssec, pp. 249–261.

[AE08b] APPEL S., EBINGER P.: *Lokalisierung in mobilen Ad-hoc-Netzen ohne zusätzliche Infrastruktur: Entfernungsschätzungen und Topologiebestimmung*. VDM Verlag Dr. Müller, Saarbrücken, 2008.

[AEGPS10] ASCHENBRUCK N., ERNST R., GERHARDS-PADILLA E., SCHWAMBORN M.: BonnMotion: A Mobility Scenario Generation and Analysis Tool. In *Proceedings of the 3rd International ICST Conference on Simulation Tools and Techniques* (2010).

[AFMT04] ASCHENBRUCK N., FRANK M., MARTINI P., TÖLLE J.: Human Mobility in MANET Disaster Area Simulation – A Realistic Approach. In *Proceedings of the 29th Annual IEEE International Conference on Local*

Computer Networks (LCN) (2004), IEEE, pp. 668–675.

[AG98] ARTMAN H., GARBIS C.: Situation awareness as distributed cognition. In *Proceedings of the European Conference on Cognitive Ergonomics 1998 (ECCE'98)* (1998).

[AGOR02] ARULAMPALAM M., GORDON N., ORTON M., RISTIC B.: A variable structure multiple model particle filter for GMTI tracking. In *Information Fusion, 2002. Proceedings of the Fifth International Conference on* (July 2002), vol. 2, pp. 927–934.

[AGPM08] ASCHENBRUCK N., GERHARDS-PADILLA E., MARTINI P.: A survey on mobility models for performance analysis in tactical mobile networks. *Journal of Telecommunication and Information Technology 2* (2008), 54–61.

[AMC11] ASCHENBRUCK N., MUNJAL A., CAMP T.: Trace-based mobility modeling for multi-hop wireless networks. *Computer Communications 34* (May 2011), 704–714.

[AO05] ALI A. A., OMAR A.: Time of arrival estimation for WLAN indoor positioning systems using matrix pencil super resolution algorithm. In *Proceedings of the 2nd Workshop on Positioning, Navigation and Communication (WPNC 05) and 1st Ultra-Wideband Expert Talk (UET 05)* (2005).

[App06] APPEL S.: *Lokalisierung von Knoten in Mobilen Ad-hoc-Netzen ohne zusätzliche Infrastruktur.* Bachelor's thesis, Technische Universität Darmstadt, Darmstadt, Germany, 2006.

[ARH00] ABDUL-RAHMAN A., HAILES S.: Supporting Trust in Virtual Communities. In *Proceedings of the 33rd Hawaii International Conference on System Sciences (HICSS)* (2000), vol. 6, IEEE.

[ARY95] ANDERSEN J. B., RAPPAPORT T. S., YOSHIDA S.: Propagation Measurements and Models for Wireless Communications Channels. *IEEE Communications Magazine 33* (Jan. 1995), 42–49.

[AS03] AGATE C. S., SULLIVAN K. J.: Road-Constrained Target Tracking and Identification Using a Particle Filter. In *Proceedings of SPIE 5204, Signal and Data Processing of Small Targets* (2003), pp. 532–543.

[Bal02] BALANIS C. A.: *Advanced Engineering Electromagnetics*, 2nd ed. Hamilton Printing Company, 2002.

[Bar02] BARDWELL J.: Converting Signal Strength Percentage to dBm Values, Nov. 2002. Available online at `http://madwifi-project.org/attachment/wiki/UserDocs/RSSI/Converting_Signal_Strength.pdf`; accessed 2013-06-17.

[Bar04a] BARR R.: *An Efficient, Unifying Approach to Simulation Using Virtual Machines*. PhD thesis, Cornell University, 2004.

[Bar04b] BARR R.: JiST: Embedding Simulation Time into a Virtual Machine. In *Proceedings of the EuroSim Congress on Modelling and Simulation* (2004).

[Bar04c] BARR R.: *JiST Java in Simulation Time, User Guide*, 2004.

[Bar04d] BARR R.: *SWANS Scalable Wireless Ad hoc Network Simulator, User Guide*, 2004.

[BAR12] BERNTORP K., ARZEN K., ROBERTSSON A.: Storage efficient particle filters with multiple out-of-sequence measurements. In *Information Fusion (FUSION), 2012 15th International Conference on* (July 2012), pp. 471–478.

[Bas00] BASS T.: Intrusion detection systems and multisensor data fusion. *Communications of the ACM 43*, 4 (2000), 99–105.

[BB02a] BUCHEGGER S., BOUDEC J.-Y. L.: Nodes Bearing Grudges: Towards Routing Security, Fairness, and Robustness in Mobile Ad Hoc Networks. In *Proceedings of the 10th Euromicro Workshop on Parallel, Distributed and Network-based Processing* (Jan. 2002), IEEE, pp. 403–410.

[BB02b] BUCHEGGER S., BOUDEC J.-Y. L.: Performance Analysis of the CONFIDANT Protocol. In *Proceedings of the 3rd ACM International Symposium on Mobile Ad Hoc Networking & Computing* (June 2002), ACM, pp. 226–236.

[BB03a] BANSAL S., BAKER M.: *Observation-based Cooperation Enforcement in Ad hoc Networks*. Tech. rep., Stanford University, July 2003.

[BB03b] BUCHEGGER S., BOUDEC J.-Y. L.: *Coping with False Accusations in Misbehavior Reputation Systems for Mobile Ad-hoc Networks.* Tech. rep., École Polytechnique Fédéale de Lausanne (EPFL), 2003.

[BB04] BUCHEGGER S., BOUDEC J.-Y. L.: A Robust Reputation System for P2P and Mobile Ad-hoc Networks. In *Second Workshop on the Economics of Peer-to-Peer Systems* (2004).

[BB06] BADIA L., BUI N.: A group mobility model based on nodes' attraction for next generation wireless networks. In *Proceedings of the 3rd international conference on Mobile technology, applications & systems* (2006), ACM.

[BBK94] BETH T., BORCHERDING M., KLEIN B.: Valuation of Trust in Open Networks. In *Proceedings of the Third European Symposium on Research in Computer Security* (1994), Springer, pp. 3–18.

[Ber04] BERIZZI A.: The Italian 2003 blackout. In *Power Engineering Society General Meeting* (2004), IEEE.

[BETJ06] BUCHER T., EBINGER P., TÖLLE J., JAHNKE M.: Modellierung und Analyse von Angriffen auf das MANET-Routing in OLSR. In *D-A-CH Mobility 2006* (2006), Horster P., (Ed.), syssec, pp. 187–200.

[Biß08] BISSMEYER N.: *Distributed Data Collection and Analysis for Attack Detection in Mobile Ad hoc Networks.* Master's thesis, FH Joanneum, Kapfenberg, Austria, 2008.

[BL04] BLAKELY K., LOWEKAMP B.: A structured group mobility model for the simulation of mobile ad hoc networks. In *MobiWac '04: Proceedings of the second international workshop on Mobility management & wireless access protocols* (2004), ACM, pp. 111–118.

[BL08] BAGGIO A., LANGENDOEN K.: Monte Carlo Localization for Mobile Wireless Sensor Networks. *Ad Hoc Networks 6*, 5 (July 2008), 718–733.

[Bla09] BLANCHARD J.: When the grid goes out: Backup power in disaster areas. In *Wireless Communication, Vehicular Technology, Information Theory and Aerospace Electronic Systems Technology, 2009. Wireless VITAE 2009. 1st International Conference on* (2009).

[BLB03] BUCHEGGER S., LE BOUDEC J.-Y.: The Effect of Rumor Spreading in Reputation Systems for Mobile Ad-hoc Networks. In *WiOpt '03: Modeling and Optimization in Mobile, Ad Hoc and Wireless Networks* (2003).

[BMB08] BUCHEGGER S., MUNDINGER J., BOUDEC J.-Y. L.: Reputation Systems for Self-Organized Networks. In *IEEE Technology and Society Magazine* (2008).

[BOP06] BARSOCCHI P., OLIGERI G., POTORTÌ F.: *Validation for 802.11b Wireless Channel Measurements.* Tech. rep., Institute of Information Science and Technologies (ISTI) in Pisa, Italian National Research Council (CNR), 2006.

[BP02] BLASCH E. P., PLANO S.: JDL level 5 fusion model: user refinement issues and applications in group tracking. In *Proceedings of the SPIE 2002* (2002), pp. 270–279.

[BPPR09] BACHRACH Y., PARNES A., PROCACCIA A. D., ROSENSCHEIN J. S.: Gossip-based aggregation of trust in decentralized reputation systems. *Autonomous Agents and MultiAgent Systems 19* (2009), 153–172.

[BQX08] BISWAS P. K., QI H., XU Y.: Mobile-agent-based collaborative sensor fusion. *Information Fusion 9*, 3 (2008), 399–411.

[Bra89] BRADEN R.: Requirements for Internet Hosts – Communication Layers. RFC 1122, 1989.

[Bra11] BRANIGAN T.: Tsunami, Earthquake, Nuclear Crisis – Now Japan Faces Power Cuts. The Guardian (London), Mar. 2011. Available online at `http://www.guardian.co.uk/world/2011/mar/13/japan-tsunami-earthquake-power-cuts`; accessed 2013-06-17.

[Buc04] BUCHEGGER S.: *Coping with Misbehavior in Mobile Ad-hoc Networks.* PhD thesis, École Polytechnique Fédérale de Lausanne (EPFL), Lausanne, Switzerland, 2004.

[Byt05] BYTOMSKI P.: SEEBA: Schnelle Einsatz Einheit Bergung Ausland, June 2005. Presentation of German Federal Agency for Technical Relief (THW).

[BZC11] BALDI M., ZANAJ E., CHIARALUCE F.: Performance Evaluation of Some Distributed Averaging Algorithms for Sensor Networks. *International Journal of Distributed Sensor Networks 2011* (2011).

[CAG12] CHOOBINEH J., ANDERSON E. E., GRIMAILA M. R.: Measuring Impact on Missions and Processes: Assessment of Cyber Breaches. In *45th Hawaii International Conference on System Science (HICSS)* (Jan. 2012), pp. 3307–3316.

[CBD02] CAMP T., BOLENG J., DAVIES V.: A Survey of Mobility Models for Ad Hoc Network Research. *Wireless Communications and Mobile Computing 2*, 5 (2002), 483–502.

[CBS03] CHRISTIAN BETTSTETTER G. R., SANTI P.: The Node Distribution of the Random Waypoint Mobility Model for Wireless Ad Hoc Networks. *IEEE Transactions on Mobile Computing 2*, 3 (2003), 257–269.

[CGM11a] CHENG N., GOVINDAN K., MOHAPATRA P.: Exploiting Mobility for Trust Propagation in Mobile Ad Hoc Networks. In *Computer Communications and Networks (ICCCN), 2011 Proceedings of 20th International Conference on* (Aug. 2011), pp. 1–6.

[CGM11b] CHENG N., GOVINDAN K., MOHAPATRA P.: Rendezvous Based Trust Propagation to Enhance Distributed Network Security. *International Journal of Security and Network 6*, 2/3 (2011), 112–122.

[Cha08] CHANG K.: Analytical Performance Evaluation for Autonomous Sensor Fusion. In *Proceedings of SPIE Defense and Security Symposium* (2008).

[CJ03] CLAUSEN T., JACQUET P.: Optimized Link State Routing Protocol (OLSR). RFC 3626 (Experimental), Oct. 2003.

[CJI02] COMMERCE B. E., JØSANG A., ISMAIL R.: The Beta Reputation System. In *Proceedings of the 15th Bled Electronic Commerce Conference* (2002).

[Cou05] COUTURIER M.: *OASIS User Requirements synthesis*. Tech. rep., OASIS Consortium, 2005.

[Cou08] COUTURIER M.: *Information security for field workers in crisis situations.* Tech. rep., OASIS Consortium, 2008.

[CSC11] CHO J.-H., SWAMI A., CHEN I.-R.: A Survey on Trust Management for Mobile Ad Hoc Networks. *Communications Surveys Tutorials, IEEE 13*, 4 (2011), 562–583.

[CW09] CAICAI G., WEI C.: Ground moving target tracking with VS-IMM particle filter based on road information. In *Radar Conference, 2009 IET International* (Apr. 2009), pp. 1–4.

[DFG01] DOUCET A., FREITAS N. D., GORDON N. (Eds.): *Sequential Monte Carlo Methods in Practice.* Springer, 2001.

[DLNP05] DAS S., LAWLESS D., NG B., PFEFFER A.: Factored particle filtering for data fusion and situation assessment in urban environments. In *Proceedings of 8th International Conference on Information Fusion* (2005).

[DWS*04] DEARDEN R., WILLEKE T., SIMMONS R., VERMA V., HUTTER F., THRUN S.: Real-time fault detection and situational awareness for rovers: report on the Mars technology program task. In *Proceedings of IEEE Aerospace Conference* (2004), vol. 2, pp. 826–840.

[EB06] EBINGER P., BUCHER T.: Modelling and analysis of attacks on the MANET routing in AODV. In *Proceedings of the 5th international conference on Ad-Hoc, Mobile, and Wireless Networks* (2006), ADHOC-NOW'06, Springer, pp. 294–307.

[EB09] EBINGER P., BISSMEYER N.: TEREC: Trust Evaluation and Reputation Exchange for Cooperative Intrusion Detection in MANETs. In *Communication Networks and Services Research Conference, 2009. CNSR '09. Seventh Annual* (2009), pp. 378–385.

[EC08] ENDSLEY M. R., CONNORS E. S.: Situation awareness: State of the art. In *Power and Energy Society General Meeting – Conversion and Delivery of Electrical Energy in the 21st Century, 2008 IEEE* (July 2008), pp. 1–4.

[EHK*07] EBINGER P., HOLLICK M., KÖNIG A., KROP T., PETERS J.: *Available and Reliable Communication in Mobile Ad hoc Networks.* Tech. Rep. Deliverable PE6, SicAri project – A security architecture and its tools for ubiquitous Internet usage, June 2007.

[EHKP06] EBINGER P., HOLLICK M., KÖNIG A., PETERS J.: *Secure Routing Mechanisms.* Tech. Rep. Deliverable PE7, SicAri project – A security architecture and its tools for ubiquitous Internet usage, June 2006.

[EJ97] ENDSLEY M., JONES W. M.: *Situation Awareness Information Dominance & Information Warfare.* Tech. Rep. 97-01, Logicon Technical Services, 1997.

[EJ01] ENDSLEY M. R., JONES D.: Disruptions, interruptions, and information attack: Impact on situation awareness and decision-making. In *Proceedings of the Human Factors and Ergonomics Society 45th Annual Meeting (HFES'01)* (2001).

[EKW12] EBINGER P., KUIJPER A., WOLTHUSEN S. D.: Enhanced location tracking for tactical MANETs based on particle filters and additional information sources. In *Military Communications and Information Systems Conference (MCC)* (Oct. 2012), pp. 1–8.

[End95] ENDSLEY M. R.: Toward a theory of situation awareness in dynamic systems. *Human Factors: The Journal of the Human Factors and Ergonomics Society vol. 37* (1995), 32–64.

[End00] ENDSLEY M. R.: Theoretical underpinnings of situation awareness: a critical review. In *Situation Awareness Analysis and Measurement,* Endsley M. R., Garland D. J., (Eds.). Lawrence Erlbaum Associates, Mahwah, NJ, USA, 2000.

[End08] ENDLER S.: *Konzeption und Implementierung von Angreifermodellen und verbesserten Methoden für die Angriffserkennung in mobilen Adhoc-Netzen.* Diplom thesis, Technische Universität Darmstadt, Darmstadt, Germany, 2008.

[EP09] EBINGER P., PARSONS M.: Measuring the impact of attacks on the performance of mobile ad hoc networks. In *Proceedings of the 6th ACM symposium on Performance evaluation of wireless ad hoc, sensor, and ubiquitous networks* (2009), PE-WASUN '09, ACM, pp. 163–164.

[ES08] EBINGER P., SOMMER M.: Using Localization Information for Attack Detection in Mobile Ad hoc Networks. In *Sicherheit 2008* (2008),

Lecture Notes in Informatics (LNI), Gesellschaft für Informatik (GI), pp. 397–406.

[Esr98] ESRI: *ESRI Shapefile Technical Description.* Tech. rep., Environmental Systems Research Institute, 1998. Available online at `http://www.esri.com/library/whitepapers/pdfs/shapefile.pdf`; accessed 2013-06-17.

[Eur09] EUROPEAN COMMITTEE FOR STANDARDIZATION (CEN): CWA 15931-1 Disaster and emergency management – Shared situation awareness – Part 1: Message structure. CEN Workshop Agreement, Feb. 2009.

[EW09] EBINGER P., WOLTHUSEN S. D.: Efficient State Estimation and Byzantine Behavior Identification in Tactical MANETs. In *Proceedings of the IEEE Military Communications Conference (MILCOM)* (Oct. 2009), IEEE, pp. 1–7.

[FGA08] FOUKALAS F., GAZIS V., ALONISTIOTI N.: Cross-layer design proposals for wireless mobile networks: a survey and taxonomy. *IEEE Communications Surveys & Tutorials 10*, 1 (Jan. 2008), 70–85.

[FGR*07] FRIEDMAN R., GAVIDIA D., RODRIGUES L., VIANA A. C., VOULGARIS S.: Gossiping on MANETs: the beauty and the beast. *SIGOPS Operating Systems Review 41*, 5 (Oct. 2007), 67–74.

[FKMR09] FRIEDMAN R., KERMARREC A.-M., MIRANDA H., RODRIGUES L.: Gossip-Based Dissemination. In *Middleware for Network Eccentric and Mobile Applications.* Springer, 2009.

[Fun04] FUNIAK S.: *State Estimation of Probabilistic Hybrid Systems with Particle Filters.* PhD thesis, Department of Electrical Engineering and Computer Science, Massachusetts Institute of Technology (MIT), Cambridge, MA, USA, 2004.

[FV11] FALL K., VARADHAN K.: *The ns Manual.* UC Berkeley, LBL, USC/ISI, and Xerox PARC, Nov. 2011. Available online at `http://www.isi.edu/nsnam/ns/ns-documentation.html`; accessed 2013-06-17.

[Gam88] GAMBETTA D.: *Trust: Making and Breaking Cooperative Relationships.* Basil Blackwell, 1988, ch. Can We Trust Trust?, pp. 213–237.

[GH04] GÜNTHER A., HOENE C.: *Measuring Round trip Times to Determine the Distance between WLAN Nodes.* Tech. Rep. TKN-04-016, Telecommunication Networks Group, Technische Universität Berlin, 2004.

[GH05] GÜNTHER A., HOENE C.: Measuring round trip times to determine the distance between WLAN nodes. In *Proceedings of the 4th IFIP-TC6 international conference on Networking Technologies, Services, and Protocols; Performance of Computer and Communication Networks; Mobile and Wireless Communication Systems* (2005), NETWORKING'05, Springer, pp. 768–779.

[GH08] GUO W., HUANG X.: Mobility Model and Relay Management for Disaster Area Wireless Networks. In *Proceedings of the Third International Conference on Wireless Algorithms, Systems, and Applications (WASA)* (2008), Springer, pp. 274–285.

[GM12] GOVINDAN K., MOHAPATRA P.: Trust computations and trust dynamics in mobile adhoc networks: A survey. *IEEE Communications Surveys and Tutorials 14*, 2 (2012), 279–298.

[Gre11] GREENFIELD L.: *What Data Errors You May Find When Building A Data Warehouse.* Tech. rep., LGI Systems Incorporated, 2011. Available online at `http://www.dwinfocenter.org/errors.html`; accessed 2013-06-17.

[GS04] GANERIWAL S., SRIVASTAVA M. B.: Reputation-Based Framework for High Integrity Sensor Networks. In *Proceedings of the 2nd ACM Workshop on the Security of Ad Hoc and Sensor Networks* (Oct. 2004), ACM, pp. 66–77.

[Guo12] GUO S.: Performance analysis of wireless intruder geolocation in campus wireless networks. In *Proceedings of the IEEE Military Communications Conference (MILCOM)* (2012).

[GV] GREIS M., VINT GROUP: *Marc Greis' Tutorial for the UCB/LBNL/VINT Network Simulator ns.* Available online at `http://www.isi.edu/nsnam/ns/tutorial/index.html`; accessed 2013-06-17.

[GW98] GENG N., WIESBECK W.: *Planungsmethoden für die Mobilkommunikation*. Springer, 1998.

[GZC*10] GEORGE S., ZHOU W., CHENJI H., WON M., LEE Y. O., PAZARLOGLOU A., STOLERU R., BAROOAH P.: DistressNet: a wireless ad hoc and sensor network architecture for situation management in disaster response. *IEEE Communications Magazine 48*, 3 (Mar. 2010), 128–136.

[GZH*12] GE L., ZHANG D., HARDY R., LIU H., YU W., RESCHLY R.: On effective sampling techniques for host-based intrusion detection in MANET. In *Proceedings of the IEEE Military Communications Conference (MILCOM)* (2012).

[Har05] HARDING L.: Snow brings chaos to Europe. The Guardian, Nov. 2005. Available online at `http://www.guardian.co.uk/environment/2005/nov/28/weather.climatechange`; accessed 2013-06-17.

[HC07] HUSSAIN F. K., CHANG E.: An Overview of the Interpretations of Trust and Reputation. In *AICT '07: Proceedings of the The Third Advanced International Conference on Telecommunications* (2007).

[HE04] HU L., EVANS D.: Localization for Mobile Sensor Networks. In *10th ACM Annual International Conference on Mobile Computing and Networking (MobiCom)* (2004).

[HGPC99] HONG X., GERLA M., PEI G., CHIANG C.-C.: A Group Mobility Model for Ad Hoc Wireless Networks. In *2nd ACM International Conference on Modeling, Analysis and Simulation of Wireless and Mobile Systems (MSWiM '99)* (1999).

[HHL88] HEDETNIEMI S. T., HEDETNIEMI S. M., LIESTMAN A.: A survey of gossiping and broadcasting in communication networks. *Networks 18* (1988), 129–134.

[HL01] HALL D. L., LLINAS J.: *Handbook of Multisensor Data Fusion*, 1st ed. CRC Press, 2001.

[HL03] HUANG Y., LEE W.: A Cooperative Intrusion Detection System for Ad Hoc Networks. In *Proceedings of the 1st ACM workshop on Security of Ad Hoc and Sensor Networks* (Oct. 2003), ACM, pp. 135–147.

[HM04] Hall D. L., McMullen S. A.: *Mathematical Techniques in Multisensor Data Fusion*, 2nd ed. Arctech House, 2004.

[HPJ03] Hu Y.-C., Perrig A., Johnson D. B.: Packet Leashes: A Defense against Wormhole Attacks in Wireless Networks. In *Proceedings of IEEE Infocomm 2003* (2003), IEEE Infocom.

[Hun03] Hunter A.: Evaluating significance of inconsistencies. In *Proceedings of the 18th international joint conference on Artificial intelligence* (2003), pp. 468–473.

[HW02] Hofbaur M. W., Williams B. C.: Mode Estimation of Probabilistic Hybrid Systems. In *Proceedings of the 5th International Workshop on Hybrid Systems: Computation and Control (HSCC)* (2002), Springer, pp. 253–266.

[HW04] Hofbaur M. W., Williams B. C.: Hybrid estimation of complex systems. *IEEE Transactions on Systems, Man, and Cybernetics – Part B: Cybernetics 34* (2004), 2178–2191.

[HWL99] Hoppe R., Wölfle G., Landstorfer F. M.: Fast 3-D Ray Tracing for the Planning of Microcells by Intelligent Preprocessing of the Database. In *Proceedings of the 3rd European Personal and Mobile Communications Conference (EPMCC '99)* (Mar. 1999), pp. 149–154.

[HWLC04] Hofmann-Wellenhof B., Lichtenegger H., Collins J.: *Global Positioning System: Theory and Practice.* Springer, 2004.

[HZ07] Huang R., Zaruba G. V.: Incorporating Data from Multiple Sensors for Localizing Nodes in Mobile Ad Hoc Networks. *IEEE Transactions on Mobile Computing 6* (Sept. 2007), 1090–1104.

[IRHS12] Islam Z., Romanuik M., Heydari S., Salmanian M.: OLSR-based coarse localization in tactical MANET situational awareness systems. In *Proceedings of IEEE International Conference on Communications (ICC)* (2012).

[ISO94] ISO: *ISO/IEC 7498-1:1994: Information Technology – Open Systems Interconnection – Basic Reference Model: The Basic Model.* Tech. rep., International Organization for Standardization, 1994.

[Jak12] JAKOBSON G.: Using federated adaptable multi-agent systems in achieving cyber attack tolerant missions. In *Cognitive Methods in Situation Awareness and Decision Support (CogSIMA), 2012 IEEE International Multi-Disciplinary Conference on* (Mar. 2012), pp. 96–102.

[JBRAS03] JARDOSH A., BELDING-ROYER E. M., ALMEROTH K. C., SURI S.: Towards realistic mobility models for mobile ad hoc networks. In *Proceedings of the 9th annual international conference on Mobile computing and networking (MobiCom '03)* (2003), ACM, pp. 217–229.

[JBRAS05] JARDOSH A. P., BELDING-ROYER E. M., ALMEROTH K. C., SURI S.: Real-world Environment Models For Mobile Network Evaluation. *IEEE Journal on Selected Areas in Communications 23*, 3 (Mar. 2005), 622–632.

[JM96] JOHNSON D. B., MALTZ D. A.: Dynamic source routing in ad hoc wireless networks. *Mobile Computing 353* (1996), 153–181.

[JWK*08] JAHNKE M., WENZEL A., KLEIN G., ASCHENBRUCK N., GERHARDS-PADILLA E., EBINGER P., KARSCH S., HAAG J., ZAEFFERER F., FINKENBRINK A.: *Abschlussbericht MITE Phase II: Dokumentation zum Forschungsvorhaben MITE (MANET Intrusion Detection for Tactical Environments) Phase II.* Tech. rep., FGAN, Universität Bonn, Fraunhofer IGD, and Fachhochschule Köln, 2008.

[Kal60] KALMAN R. E.: A New Approach to Linear Filtering and Prediction Problems. *Transactions of the ASME–Journal of Basic Engineering 82*, Series D (1960), 35–45.

[Kaw12] KAWAMOTO K.: A particle filter with optimal discrete density for hybrid state estimation. In *Procedding of International Symposium on Communications and Information Technologies (ISCIT)* (2012), pp. 296–299.

[KB61] KALMAN R. E., BUCY R. S.: New results in linear filtering and prediction theory. *Transactions of the ASME–Journal of Basic Engineering 83*, Series D (1961), 95–107.

[KBHS07] KROP T., BREDEL M., HOLLICK M., STEINMETZ R.: JiST/MobNet: combined simulation, emulation, and real-world testbed for ad hoc networks. In *WinTECH '07: Proceedings of the the second ACM international workshop on Wireless network testbeds, experimental evaluation and characterization* (2007), ACM, pp. 27–34.

[KBR97] KRIZMAN K., BIEDKA T., RAPPAPORT T.: Wireless position location: Fundamentals, implementation strategies, and sources of error. In *IEEE VTC 97* (1997), pp. 919–923.

[KCMC05] KURKOWSKI S., CAMP T., MUSHELL N., COLAGROSSO M.: A visualization and analysis tool for NS-2 wireless simulations: iNSpect. In *Modeling, Analysis, and Simulation of Computer and Telecommunication Systems, 2005. 13th IEEE International Symposium on* (Sept. 2005), pp. 503–506.

[KDG03] KEMPE D., DOBRA A., GEHRKE J.: Gossip-Based Computation of Aggregate Information. In *44th Symposium on Foundations of Computer Science (FOCS 2003)* (2003), IEEE, pp. 482–491.

[KG03] KACHIRSKI O., GUHA R.: Effective Intrusion Detection Using Multiple Sensors in Wireless Ad Hoc Networks. In *Proceedings of the 36th Annual Hawaii International Conference on System Sciences (HICSS'03), Track 2* (2003), vol. 2, IEEE.

[KHBV*04] KEAN T. H., HAMILTON L. H., BEN-VENISTE R., FIELDING F. F., GORELICK J. S., GORTON S., KERREY B., LEHMAN J. F., ROEMER T. J., THOMPSON J. R.: *The 9/11 Commission Report: Final Report of the National Commission on Terrorist Attacks Upon the United States.* W. W. Norton& Company, 2004.

[KK05] KAWADIA V., KUMAR P. R.: A cautionary perspective on cross-layer design. *IEEE Wireless Communications 12*, 1 (Feb. 2005), 3–11.

[KL92] KIFER M., LOZINSKII E. L.: A Logic for Reasoning with Inconsistency. *Automated Reasoning 9* (Oct. 1992), 179–215.

[KLAML11] KIDSTON D., LI L., AL MAMUN W., LUTFIYYA H.: Cross-layer cluster-based data dissemination for failure detection in MANETs. In *Proceedings of the 7th International Conference on Network and Services Management (CNSM '11)* (2011), pp. 267–273.

[KLTM10] KIDSTON D., LI L., TANG H., MASON P.: Mitigating Security Threats in Tactical Networks. In *Proceedings of Information Systems and Technology Panel (IST) Symposium* (2010).

[Kön06] KÖNIG A.: Geographically Secure Routing for Mobile Ad Hoc Networks: A Cross-Layer Based Approach. In *Proceedings of the International Conference on Emerging Trends in Information and Communication Security (ETRICS)* (2006), Müller G., (Ed.), LNCS, Springer.

[Kni02] KNIGHT K.: Measuring Inconsistency. *Journal of Philosophical Logic 31* (Feb. 2002), 77–98.

[KPCD12] KAN Z., PASILIAO E., CURTIS J., DIXON W.: Particle filter based average consensus target tracking with preservation of network connectivity. In *Proceedings of the IEEE Military Communications Conference (MILCOM)* (2012).

[KSGM03] KAMVAR S. D., SCHLOSSER M. T., GARCIA-MOLINA H.: The Eigentrust algorithm for reputation management in P2P networks. In *Proceedings of the 12th international conference on World Wide Web (WWW '03)* (2003), ACM, pp. 640–651.

[Lam78] LAMPORT L.: Time, clocks, and the ordering of events in a distributed system. *Communications of the ACM 21*, 7 (July 1978), 558–565.

[LCM*03] LANGE C., CHAUDRON M. R. V., MUSKENS J., SOMERS L. J., DORTMANS H. M.: An Empirical Investigation in Quantifying Inconsistency and Incompleteness of UML Designs. In *Incompleteness of UML Designs, Proceedings of Workshop on Consistency Problems in UML-based Software Development, 6th International Conference on Unified Modeling Language, UML 2003* (2003).

[LJT04] LIU Z., JOY A. W., THOMPSON R. A.: A Dynamic Trust Model for Mobile Ad Hoc Networks. *IEEE International Workshop on Future Trends of Distributed Computing Systems 0* (2004), 80–85.

[LMSK06] LEINMÜLLER T., MAIHÖFER C., SCHOCH E., KARGL F.: Improved security in geographic ad hoc routing through autonomous position verification. In *VANET '06: Proceedings of the 3rd international workshop on Vehicular ad hoc networks* (2006), ACM, pp. 57–66.

[LPM*05] LAZOS L., POOVENDRAN R., MEADOWS C., SYVERSON P., CHANG L. W.: Preventing Wormhole Attacks on Wireless Ad Hoc Networks: A Graph Theoretic Approach. In *IEEE Wireless Communications and Networking Conference (WCNC)* (2005).

[LSK06] LEINMÜLLER T., SCHOCH E., KARGL F.: Position Verification Approaches for Vehicular Ad Hoc Networks. *IEEE Wireless Communications, Special Issue on Inter-Vehicular Communications 13*, 5 (Oct. 2006), 16–21.

[LSKM10] LEINMÜLLER T., SCHOCH E., KARGL F., MAIHÖFER C.: Decentralized position verification in geographic ad hoc routing. *Security and Communication Networks 3*, 4 (July 2010), 289–302.

[LW07] LI F., WU J.: *Mobility Reduces Uncertainty in MANETs*. Tech. rep., Department of Computer Science and Engineering, Florida Atlantic University, Boca Raton, FL 33431, 2007.

[LY03] LIU Y., YANG Y. R.: Reputation propagation and agreement in mobile ad-hoc networks. In *Wireless Communications and Networking, 2003. WCNC 2003. 2003 IEEE* (Mar. 2003), vol. 3, pp. 1510–1515.

[Mau96] MAURER U. M.: Modelling a Public-Key Infrastructure. In *ESORICS '96: Proceedings of the 4th European Symposium on Research in Computer Security* (1996), pp. 325–350.

[Mau05] MAURER J.: *Strahlenoptisches Kanalmodell für die Fahrzeug-Fahrzeug-Funkkommunikation*. PhD thesis, Institut fuer Höchstfrequenztechnik und Elektronik, Universität Karlsruhe, Karlsruhe, Germany, July 2005.

[MEKB06] MAUTHNER M., ELMENREICH W., KIRCHNER A., BOESEL D.: Out-of-Sequence Measurement Treatment in Sensor Fusion Applications: Buffering versus Advanced Algorithms. In *Workshop Fahrerassistenzsysteme* (2006), pp. 20–30.

[MFSW04] MAURER J., FÜGEN T., SCHÄFER T., WIESBECK W.: A New Inter-Vehicle Communications (IVC) Channel Model. In *Proceedigns of the 60th IEEE Vehicular Technology Conference (VTC2004-Fall)* (2004).

[MFW05] MAURER J., FÜGEN T., WIESBECK W.: Physical layer simulations of IEEE802.11a for vehicle-to-vehicle communications. In *IEEE 62nd Vehicular Technology Conference* (2005).

[MGLB00] MARTI S., GIULI T. J., LAI K., BAKER M.: Mitigating Routing Misbehavior in Mobile Ad Hoc Networks. In *Proceedings of the 6th Annual International Conference on Mobile computing and Networking* (Aug. 2000), ACM, pp. 255–265.

[Mic11] MICROSOFT DEVELOPER NETWORK (MSDN): Visual Studio .NET 2003: Data Validation, 2011. Available online at `http://msdn.microsoft.com/en-us/library/aa291820%28v=vs.71%29.aspx`; accessed 2013-06-17.

[MM02] MICHIARDI P., MOLVA R.: Core: A Collaborative Reputation Mechanism to Enforce Node Cooperation in Mobile Ad Hoc Networks. In *Proceedings of the IFIP TC6/TC11 Sixth Joint Working Conference on Communications and Multimedia Security* (2002), Kluwer Academic Publishers, pp. 107–121.

[MMH02] MUI L., MOHTASHEMI M., HALBERSTADT A.: A Computational Model of Trust and Reputation. In *Proceedings of the 35th Annual Hawaii International Conference on System Sciences (HICSS'02)* (2002), vol. 7, IEEE, pp. 188–196.

[MNP04] MISHRA A., NADKARNI K., PATCHA A.: Intrusion detection in wireless ad hoc networks. *Wireless Communications 11* (2004), 48–60.

[MR03] MIRANDA H., RODRIGUES L.: Friends and foes: preventing selfishness in open mobile ad hoc networks. In *Distributed Computing Systems Workshops, 2003. Proceedings. 23rd International Conference on* (May 2003), pp. 440–445.

[MR04] MOOKERJEE P., REIFLER F.: Application of reduced state estimation to multisensor fusion with out-of-sequence measurements. In *Radar Conference, 2004. Proceedings of the IEEE* (Apr. 2004), pp. 111–116.

[MU08a] MERTENS M., ULMKE M.: Ground Moving Target Tracking with context information and a refined sensor model. In *Information Fusion, 2008 11th International Conference on* (July 2008), pp. 1–8.

[MU08b] MERTENS M., ULMKE M.: Precision GMTI tracking using road constraints with visibility information and a refined sensor model. In *Radar Conference, 2008. RADAR '08. IEEE* (May 2008), pp. 1–6.

[Nat05] The Deadliest Tsunami in History? National Geographic News, Jan. 2005. Available online at `http://news.nationalgeographic.com/news/2004/12/1227_041226_tsunami.html`; accessed 2013-06-17.

[Ngu08] NGUYEN N. T.: *Advanced Methods for Inconsistent Knowledge Management.* Springer, 2008.

[OBDWU08] ONG L.-L., BAILEY T., DURRANT-WHYTE H., UPCROFT B.: Decentralised particle filtering for multiple target tracking in wireless sensor networks. In *11th International Conference on Information Fusion* (July 2008), pp. 1–8.

[ODK11] OPITZ F., DASTNER K., KOHLER B.: Non linear techniques in GMTI tracking. In *Radar Symposium (IRS), 2011 Proceedings International* (Sept. 2011), pp. 187–192.

[OG08] ORGUNER U., GUSTAFSSON F.: Storage efficient particle filters for the out of sequence measurement problem. In *Information Fusion, 2008 11th International Conference on* (July 2008), pp. 1–8.

[OUR*05] ONG L., UPCROFT B., RIDLEY M., BAILEY T., SUKKARIEH S., DURRANT-WHYTE H.: Decentralised data fusion with particles. In *Proceedings of the 2005 Australasian Conference on Robotics & Automation (ACRA)* (2005).

[OUR*06] ONG L.-L., UPCROFT B., RIDLEY M., BAILEY T., SUKKARIEH S., DURRANT-WHYTE H.: Consistent methods for Decentralised Data Fusion using Particle Filters. In *Multisensor Fusion and Integration for Intelligent Systems, 2006 IEEE International Conference on* (Sept. 2006), pp. 85–91.

[Par08] PARSONS M.: *Evaluation der Auswirkungen verschiedener Angriffstypen in mobilen Ad-hoc-Netzen.* Bachelor's thesis, Technische Universität Darmstadt, Darmstadt, Germany, 2008.

[Paw98] PAWLAK Z.: An inquiry into anatomy of conflicts. *Information Sciences 109*, 1-4 (1998), 65–78.

[PBRD03] PERKINS C. E., BELDING-ROYER E. M., DAS S. R.: Ad hoc On-Demand Distance Vector (AODV) Routing. RFC 3561, July 2003.

[PD11] PATHAK P., DUTTA R.: A Survey of Network Design Problems and Joint Design Approaches in Wireless Mesh Networks. *Communications Surveys Tutorials, IEEE 13*, 3 (2011), 396–428.

[PDP08] PANNETIER B., DEZERT J., POLLARD E.: Improvement of Multiple Ground Targets Tracking with GMTI Sensor and Fusion of Identification Attributes. In *Aerospace Conference, 2008 IEEE* (Mar. 2008), pp. 1–13.

[PDTS*09] PAULHEIM H., DÖWELING S., TSO-SUTTER K. H. L., PROBST F., ZIEGERT T.: Improving Usability of Integrated Emergency Response Systems: The SoKNOS Approach. In *Proceedings of the Informatik 2009* (Oct. 2009), Fischer S., Maehle E., Reischuk R., (Eds.), vol. 154 of *Lecture Notes in Informatics (LNI)*, Gesellschaft für Informatik (GI), pp. 1435–1349.

[PE09] PARSONS M. J., EBINGER P.: Performance Evaluation of the Impact of Attacks on Mobile Ad hoc Networks. In *Proceedings of Workshops Dependable Network Computing and Mobile Systems; Field Failure Data Analysis; Embedded Systems and Communications Security* (2009), pp. 40–48.

[PPBS10] PROBST F., PAULHEIM H., BRUCKER A., SCHULTE S.: SoKNOS – Informationsdienste für das Katastrophenmanagement. In *Proceedings of the VDE-Kongress 2010 – E-Mobility: Technologien, Infrastruktur, Märkte* (Nov. 2010), VDE, VDE Verlag GmbH.

[RAG04] RISTIC B., ARULAMPALAM S., GORDON N.: *Beyond the Kalman Filter: Particle Filters for Tracking Applications*. Artech House, 2004.

[RAOR91] RUSTAKO A., AMITAY N., OWENS G., ROMAN R.: Radio propagation at microwave frequencies for line-of-sightmicrocellular mobile and personal communications. *IEEE Transactions on Vehicular Technology 40* (1991), 203–210.

[Rap02] RAPPAPORT T. S.: *Wireless Communications: Principles and Practice, Second Edition*, 2nd ed. Prentice Hall PTR, Englewood Cliffs, NJ, USA, 2002.

[Rei06] REIDT S.: *Topographisches Routing in mobilen Ad-hoc-Netzen*. Diplom thesis, Technische Universität Darmstadt, Darmstadt, Germany, 2006.

[REKW11] REIDT S., EBINGER P., KUIJPER A., WOLTHUSEN S.: Resource-Constrained Signal Propagation Modeling for Tactical Mobile Ad Hoc Networks. In *Proceedings of the 2011 IEEE First International Workshop on Network Science (NSW 2011)* (June 2011), pp. 67–74.

[RGS*06] RAHMATOLLAHI G., GALLER S., SCHROEDER J., JOBMANN K., KYAMAKYA K.: Propagation delay based positioning using IEEE 802.11b signals. In *Proceedigns of the 3rd Workshop on Positioning, Navigation and Communication (WPNC 06)* (2006).

[RGT03] ROSENCRANTZ M., GORDON G., THRUN S.: Decentralized Sensor Fusion with Distributed Particle Filters. In *19th Conference on Uncertainty in Artificial Intelligence (UAI)* (2003).

[RHW09] RAUTIAINEN T., HOPPE R., WÖLFLE G.: Measurements and 3D Ray Tracing Propagation Predictions of Channel Characteristics in Indoor Environments. In *Proceedings of the 18th IEEE Annual International Symposium on Personal Indoor and Mobile Radio Communications (PIMRC '09)* (Sept. 2009), IEEE Communications Society, pp. 1–5.

[RL10] ROMERO S., LACEY M.: Fierce Quake Devastates Haitian Capital. New York Times, 2010. Available online at `http://www.nytimes.com/2010/01/13/world/americas/13haiti.html?_r=0`; accessed 2013-09-21.

[Roe00] ROESSINGH J.: *ESSAI – WP1 Orientation on Situation Awareness and Crisis Management*. Tech. rep., Nationaal Lucht- en Ruimtevaartlaboratorium (NLR) on behalf of Enhanced Safety through Situation Awareness Integration in training (ESSAI) consortium, 2000.

[RWH02] RAUTIAINEN T., WOLFLE G., HOPPE R.: Verifying Path Loss and Delay Spread Predictions of a 3D Ray Tracing Propagation Model in

Urban Environments. In *Proceedings of the 2002 IEEE 56th Vehicular Technology Conference* (May 2002), IEEE Communications Society, pp. 2470–2474.

[Sal07] SALERNO J. J.: Where's level 2/3 fusion – a look back over the past 10 years. In *Information Fusion, 2007 10th International Conference on* (July 2007), pp. 1–4.

[Sal08] SALERNO J.: Measuring situation assessment performance through the activities of interest score. In *Information Fusion, 2008 11th International Conference on* (July 2008), pp. 1–8.

[SAM10] SCHWAMBORN M., ASCHENBRUCK N., MARTINI P.: A realistic trace-based mobility model for first responder scenarios. In *Proceedings of the 13th ACM international conference on Modeling, analysis, and simulation of wireless and mobile systems (MSWIM)* (2010), ACM.

[SAP08] SAP AG: *SAP NetWeaver How-To Guide: How To Use Data Services I – Data Quality Made Easy*, version 1.0 ed., Nov. 2008.

[SAP13] SAP AG: Collateral Management System: Plausibility Check. SAP Help Portal, 2013. Available online at `http://help.sap.com/saphelp_srm40/helpdata/ru/02/09113a448a4bd4b4d5b5e7eb99c3f8/frameset.htm`; accessed 2013-06-17.

[SBC*05] STERNE D., BALASUBRAMANYAM P., CARMAN D., WILSON B., TALPADE R., KO C., BALUPARI R., TSENG C.-Y., BOWEN T., LEVITT K., ROWE J.: A General Cooperative Intrusion Detection Architecture for MANETs. In *IWIA '05: Proceedings of the Third IEEE International Workshop on Information Assurance (IWIA'05)* (2005), IEEE, pp. 57–70.

[Sch06] SCHÖN T.: *Estimation of Nonlinear Dynamic Systems – Theory and Applications*. PhD thesis, Linkoping University, Linkoping, Sweden, 2006.

[SG11] SEARCH I., GROUP R. A.: *Guidelines and Methodology*. Tech. rep., Office for the Coordination of Humanitarian Affairs, United Nations, 2011.

[SH95] SMITH K., HANNOCK P. A.: Situation awareness is adaptive, externally directed consciousness. *Human Factors: The Journal of the Human Factors and Ergonomics Society 37*, 1 (1995), 137–148.

[SHB04] SALERNO J., HINMAN M., BOULWARE D.: Building a framework for situation awareness. In *Proceedings of the Seventh International Conference on Information Fusion* (2004), Svensson P., Schubert J., (Eds.), vol. I, International Society of Information Fusion, pp. 219–226.

[SHR05] STEPANOV I., HERRSCHER D., ROTHERMEL K.: On the Impact of Radio Propagation Models on MANET Simulation Results. In *Proceedings of the 7th IFIP International Conference on Mobile and Wireless Communications Networks (MWCN '05)* (Sept. 2005).

[SLLJ09] SOINI J., LINNA P., LEPPANIEMI J., JAAKKOLA H.: Toward collaborative situational awareness in a time-critical operational environment. In *Management of Engineering Technology, 2009. PICMET 2009. Portland International Conference on* (2009).

[SM05] SRIVASTAVA V., MOTANI M.: Cross-layer design: a survey and the road ahead. *IEEE Communications Magazine 43*, 12 (Dec. 2005), 112–119.

[Som07] SOMMER M.: *Auswertung von Positionsdaten, Funkcharakteristika und des Verbindungsgraphen für die Angriffserkennung in mobilen Ad-hoc-Netzen.* Diplom thesis, Technische Universität Darmstadt, Darmstadt, Germany, 2007.

[Spe98] SPECIAL MOBILE GROUP (SMG): *Universal Mobile Telecommunicatios System (UMTS) – Selection Procedures for the Choice of Radio Transmission Technologies of the UMTS.* Tech. rep., European Telecommunications Standards Institute (ETSI), 1998.

[Spi06] Mega-Blackout: Europa im Dunkeln, Politiker in Panik. Spiegel Online, Nov. 2006. Available online at `http://www.spiegel.de/panorama/mega-blackout-europa-im-dunkeln-politiker-in-\\panik-a-446581.html`; accessed 2013-06-17.

[SSH*06] STANTON N. A., STEWART R., HARRIS D., HOUGHTON R. J., MCMASTER R., SALMON P., HOYLE G., WALKER G., YOUNG M., LINSELL M., DYMOTT R., GREEN D.: Distributed situation awareness in dynamic systems: theoretical development and application of an ergonomics methodology. *Ergonomics 49* (2006), 1288–1311.

[SSW*08] SALMON P. M., STANTON N. A., WALKER G. H., BABER C., JENKINS D. P., MCMASTER R., YOUNG M. S.: What Really is Going on? Review of Situation Awareness Models for Individuals and Teams. *Theoretical Issues in Ergonomics Science 9*, 4 (2008), 297–323.

[SSWJ09] SALMON P. M., STANTON N. A., WALKER G. H., JENKINS D. P.: *Distributed Situation Awareness – Theory, Measurement and Application to Teamwork.* Ashgate Publishing, 2009.

[Sta03] STAATSKANZLEI DES LANDES SACHSEN-ANHALT: Innenminister Jeziorsky stellt Abschlussbericht zum Hochwasser 2002 im Land Sachsen-Anhalt vor. Pressemitteilung Nr.: 160/03, Apr. 2003. Available online at `http://www.asp.sachsen-anhalt.de/presseapp/data/stk/2003/160_2003.htm`; accessed 2013-06-17.

[Str08] STRELLER D.: Road map assisted ground target tracking. In *Information Fusion, 2008 11th International Conference on* (July 2008), pp. 1–7.

[Stu99] STULLER J.: Inconsistencies in Data Warehousing. In *Proceedings of the 1999 International Symposium on Database Applications in Non-Traditional Environments (DANTE '99)* (1999), pp. 43–50.

[SW91] SATTER N. B., WOODS D. D.: Situation Awareness: A Critical but Ill-Defined Phenomenon. *The International Journal of Aviation Psychology 1*, 1 (1991), 45–57.

[SWP03] SUN B., WU K., POOCH U.: Routing Anomaly Detection in Mobile Ad Hoc Networks. In *Proceedings of the 12th International Conference on Computer Communications and Networks (ICCCN 2003)* (Oct. 2003), IEEE, pp. 25–31.

[SWV*08] SCHAIRER T., WEISS C., VORST P., SOMMER J., HOENE C., ROSENSTIEL W., STRASSER W., ZELL A., CARLE G., SCHNEIDER P., WEIS-

BECKER A.: Integrated Scenario for Machine-Aided Inventory Using Ambient Sensors. In *4th European Workshop on RFID Systems and Technologies (RFID SysTech 2008)* (June 2008).

[Szt00] SZTOMPKA P.: *Trust: A Sociological Theory.* Cambridge University Press, 2000.

[TB04] THEODORAKOPOULOS G., BARAS J. S.: Trust Evaluation in Ad-Hoc Networks. In *Proceedings of the 2004 ACM Workshop on Wireless Security* (July 2004), ACM, pp. 1–10.

[TB06] THEODORAKOPOULOS G., BARAS J. S.: On trust models and trust evaluation metrics for ad hoc networks. *IEEE Journal on Selected Areas in Communications 24* (2006), 318–328.

[TC05] TSENG H. C., CULPEPPER B. J.: Sinkhole intrusion in mobile ad hoc networks: The problem and some detection indicators. *Computers & Security 24* (2005), 561–570.

[THB*02] TIAN J., HAHNER J., BECKER C., STEPANOV I., ROTHERMEL K.: Graph-based mobility model for mobile ad hoc network simulation. In *Simulation Symposium, 2002. Proceedings. 35th Annual* (Apr. 2002), pp. 337–344.

[TN04] THONABAUER G., NÖSSLINGER B.: *Guidelines on Credit Risk Management: Credit Approval Process and Credit Risk Management.* Oesterreichische Nationalbank, Dec. 2004.

[UK06] ULMKE M., KOCH W.: Road-map assisted ground moving target tracking. *Aerospace and Electronic Systems, IEEE Transactions on 42*, 4 (Oct. 2006), 1264–1274.

[UPI11] U.S. Tornado Damage Estimated in Billions. United Press International (UPI), Apr. 2011. Available online at `http://www.upi.com/Top_News/US/2011/04/29/US-tornado-damage-estimated-in-billions/UPI-15551304065800/`; accessed 2013-06-17.

[VDVM94] VIDULICH M., DOMINGUEZ C., VOGEL E., MCMILLAN G.: *Situation Awareness: Papers and Annotated Bibliography.* Tech. Rep. AL/CF-TR-1994-0085, Armstrong Laboratory, Crew Systems Directorate, 1994.

[VK04] VELAYOS H., KARLSSON G.: Limitations of range estimation in wireless LAN. In *Proceedings of the 1rst Workshop on Positioning, Navigation and Communication (WPNC 04)* (2004).

[VSH*08] VORST P., SOMMER J., HOENE C., SCHNEIDER P., WEISS C., SCHAIRER T., ROSENSTIEL W., ZELL A., CARLE G.: Indoor Positioning via Three Different RF Technologies. In *4th European Workshop on RFID Systems and Technologies (RFID SysTech 2008)* (June 2008).

[Wan06] WANG X.: Intrusion Detection Techniques in Wireless Ad Hoc Networks. In *Proceedings of the 30th Annual International Computer Software and Applications Conference (COMPSAC'06)* (2006), pp. 347–349.

[War05] WARRICK J.: Crisis Communications Remain Flawed – Despite Promises to Fix Systems, First Responders Were Still Isolated After Katrina, 2005.

[WB02] WANG K., BAOCHUN L.: Group mobility and partition prediction in wireless ad-hoc networks. In *Proceedings of IEEE International Conference on Communications 2002 (ICC 2002)* (2002), vol. 2, pp. 1017–1021.

[WB04] WANG W., BHARGAVA B.: Visualization of Wormholes in Sensor Networks. In *Proceedings of the 2004 ACM Workshop on Wireless Security* (Oct. 2004), ACM, pp. 51–60.

[WH06] WILLIAMS S. A., HUANG D.: A Group Force Mobility Model. In *9th Communications and Networking Simulation Symposium (CNS 2006)* (2006).

[Wil04] WILLIAM NAVIDI AND TRACY CAMP: Stationary Distributions for the Random Waypoint Mobility Model. *IEEE Transactions on Mobile Computing 3*, 1 (2004), 99–108.

[WV03] WANG Y., VASSILEVA J.: Trust and Reputation Model in Peer-to-Peer Networks. In *Proceedings of the 3rd International Conference on Peer-to-Peer Computing (P2P '03)* (2003), IEEE, pp. 150–.

[XL02] XIONG L., LIU L.: Building Trust in Decentralized Peer-to-Peer Electronic Communities. In *International Conference on Eletronic Commerce Research (ICECR-5)* (2002).

[XMLSR93] XIA H., MACIEL H. B. L., LINDSAY-STEWART A., ROWE R.: Radio propagation characteristics for line-of-sight microcellular and personal communications. *IEEE Transactions on Antennas and Propagation 41* (1993), 1439–1447.

[XPS11] XENAKIS C., PANOS C., STAVRAKAKIS I.: A comparative evaluation of intrusion detection architectures for mobile ad hoc networks. *Computers & Security 30*, 1 (2011), 63–80.

[YML02] YANG H., MENG X., LU S.: Self-organized network-layer security in mobile ad hoc networks. In *Proceedings of the 1st ACM workshop on Wireless security* (2002), WiSE '02, ACM, pp. 11–20.

[ZBS11] ZHANG S., BAR-SHALOM Y.: Optimal update with multiple out-of-sequence measurements. In *Proceedings SPIE 8050, Signal Processing, Sensor Fusion, and Target Recognition XX* (2011).

[ZBS12] ZHANG S., BAR-SHALOM Y.: Out-of-Sequence Measurement Processing for Particle Filter: Exact Bayesian Solution. *Aerospace and Electronic Systems, IEEE Transactions on 48*, 4 (Oct. 2012), 2818–2831.

[ZL00] ZHANG Y., LEE W.: Intrusion Detection in Wireless Ad-Hoc Networks. In *Proceedings of the 6th Annual International Conference on Mobile Computing and Networking* (Aug. 2000), ACM, pp. 275–283.

[ZL04] ZHANG Y., LEE W.: Security in Mobile Ad-hoc Networks. In *Ad hoc networks – Technologies and protocols*, Mohapatra P., Krishnamurthy S., (Eds.). Springer, 2004, pp. 249–268.

[ZLH03] ZHANG Y., LEE W., HUANG Y.-A.: Intrusion Detection Techniques for Mobile Wireless Networks. *Wireless Networks 9*, 5 (Sept. 2003), 545–556.

[ZMHT05] ZOURIDAKI C., MARK B. L., HEJMO M., THOMAS R. K.: A quantitative trust establishment framework for reliable data packet delivery in MANETs. In *SASN '05: Proceedings of the 3rd ACM workshop on Security of ad hoc and sensor networks* (2005).

[ZMHT06] ZOURIDAKI C., MARK B. L., HEJMO M., THOMAS R. K.: Robust Cooperative Trust Establishment for MANETs. In *SASN'06 October 30, 2006 Alexandria, Virginia, USA* (2006).

[ZZNJ02] ZHANG D., ZHOU L., NUNAMAKER JR J. F.: A Knowledge Management Framework for the Support of Decision Making in Humanitarian Assistance/Disaster Relief. *Knowledge and Information Systems 4* (2002), 370–385.